W0107713

ADHESION 15

This volume is based on papers presented at the 28th annual conference on Adhesion and Adhesives held at The City University, London

Previous conferences have been published under the titles of
Adhesion 1–14

ADHESION 15

Edited by

K. W. ALLEN

Adhesion Science Group,
The City University, London, UK

SPRINGER-SCIENCE+BUSINESS MEDIA, B.V.

WITH 29 TABLES AND 107 ILLUSTRATIONS

© 1991 SPRINGER SCIENCE+BUSINESS MEDIA DORDRECHT
Originally published by ELSEVIER SCIENCE PUBLISHERS LTD in 1991,
and CROWN COPYRIGHT - Chapters 9 and 12 in 1991.

British Library Cataloguing in Publication Data

Conference on Adhesion and Adhesives (28th: 1990: London,
  England)
  Adhesion 15
  1. Adhesion
  I. Title II. Allen K. W.
  541.33

ISBN 978-94-010-5714-1   ISBN 978-94-011-3854-3 (eBook)
DOI 10.1007/978-94-011-3854-3

Library of Congress CIP data applied for

No responsibility is assumed by the Publisher for any injury and/or damage to persons or
property as a matter of products liability, negligence or otherwise, or from any use or
operation of any methods, instructions or ideas contained in the material herein.

Special regulations for readers in the USA

This publication has been registered with the Copyright Clearance Center Inc. (CCC), Salem,
Massachusetts. Information can be obtained from the CCC about conditions under which
photocopies of parts of this publication may be made in the USA. All other copyright
questions, including photocopying outside the USA, should be referred to the publisher.

All rights reserved. No part of this publication may be reproduced, stored in a retrieval
system, or transmitted in any form or by any means, electronic, mechanical, photocopying,
recording, or otherwise, without the prior written permission of the publisher.

Preface

Once again, just before Easter, a goodly number of people gathered to hear a series of papers on a variety of aspects of adhesion presented by authors from a number of countries. Overall they all seemed to be as well pleased as ever with what they received.

One might have thought that by now—the twenty-eighth year—the whole range of possible topics had been scanned; but no, there are two papers which are concerned with an entirely new one. This arises from the fairly recent recognition that if society is to utilise adhesives and gain anything like their potential advantage, then there is a whole area of dissemination of information and a special sort of education which must be accomplished. To meet this need various initiatives have been undertaken, and two of these are reported within this volume—in addition to papers of the more conventional technological type.

As always, I must record my thanks to all those who make possible both these conferences and the books through which the papers are available to a wider audience. The audience, the authors and their secretaries, various people within the University, the publishers and their staff; all are essential parts of the whole. To them, each and every one, may I express my sincere appreciation and gratitude.

K. W. ALLEN

Contents

1

TESTING OF ADHESIVES - USEFUL OR NOT

ROBERT D. ADAMS
Professor of Applied Mechanics
Department of Mechanical Engineering
University of Bristol
Queens' Building, University Walk, Bristol BS8 1TR, UK

INTRODUCTION

In industry and in research laboratories, there is a variety of test methods used to indicate the 'strength' of adhesively bonded joints. The questions to be answered are 'why do we do these tests?' and 'what do the results mean?' If there is a reasonable answer to the second question, then it is worth continuing with the test programme. But all too often, the test results are functions of the test rather than the object being tested.

Any industrially-useful test should be simple to set up and carry out, be repeatable and should be representative of the loading which is likely to be experienced in the finished product. Such tests would be for, say, assessing variations in production quality by making measurements on coupons (test pieces) which had gone through the same production cycle, or for checking the quality of adhesives at goods in before commencing a production run.

There is another reason for testing adhesives: this often relates to the need to establish the mechanical or physical properties of an adhesive for design calculations. While these tests may well be more carefully designed and executed than production-oriented tests, they must be repeatable and easily interpreted.

The objective of this paper is to discuss the mechanics of some commonly used (misused?) tests and to indicate where these may have shortcomings. Lap, butt and peel joints will be featured. A fuller description of these tests is given in [1].

1

SINGLE LAP JOINT

The essence of this joint is that two strips of metal (or some other material) are overlapped and bonded. By applying tension to the strips in their plane, the bonded joint is subjected to shear. A typical joint is shown in Fig.1a together with the shear stress if the adherends (strips) are inextensible compared with the adhesive. The average shear stress τ is given by

$$\tau = P/bl$$

where P is the applied load,
$\quad\quad b$ is the width of the strips, and
$\quad\quad l$ is the overlap length

This is often referred to as tensile shear, which is a nonsensical contradiction, but is derived from the fact that *tensile* loads are used to generate *shear* stresses. This testing is standardised as ASTM D 1002-72. A practical point is that tabs should be

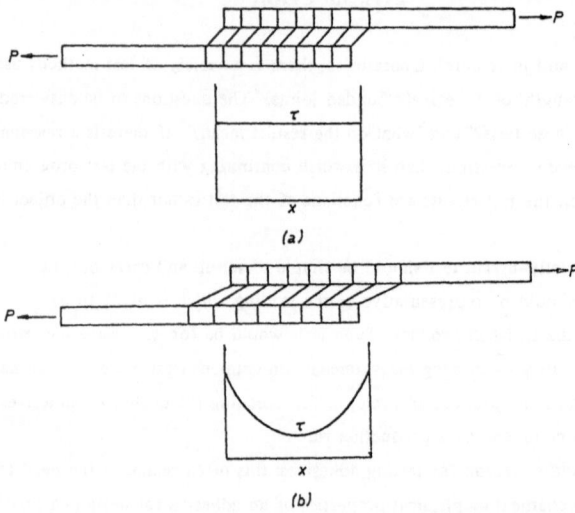

(a)

(b)

Figure 1. Exaggerated deformations in loaded single lap joint showing the adhesive shear stress, τ (a) with rigid adherends, (b) with elastic adherends

bonded to the ends of the strips to help alignment in the jaws of the testing machine used to apply the tension. In practice, the adherends stretch under the applied load, giving rise to differential shears as shown in Fig.1b.

Volkersen [2] developed a set of equations which gave the variation of shear stress with distance along the bond line (see Reference [1], p.20 *et seq*). This gives the bathtub curve shown in Fig. 1b. The shear strain (and stress) in the adhesive layer is a maximum at each end, and a minimum in the middle. If the adherends are not of equal stiffness, the adhesive shear and adherend tensile stresses will b asymmetric.

But Volkersen ignored the fact that the directions of the two forces in Fig.1 are not collinear, which creates a bending moment applied to the joint in addition to the in-plane tension. The adherends bend and the rotation alters the direction of the load line in the region of the overlap in such a way that the joint displacements are no longer directly proportional to the applied load. Goland and Reissner [3] took this effect into account by using a bending moment factor, k, which relates the bending moment on the adherend at the end of the overlap, M_0, to the in-plane loading by the relationship

$$M_0 = kPt/2$$

where t is the adherend thickness (the thickness of the adhesive layer was neglected).

If the load on the joint is very small, no rotation of the overlap takes place, so $M_0 = Pt/2$ and k = 1.0. As the load is increased, the overlap rotates, bringing the line of action of the load closer to the centre-line of the adherends, thus reducing the value of the bending moment factor. Goland and Reissner give a similar shear stress distribution to that of Volkersen, but also give the transverse (peel) stresses σ_y, in the adhesive layer.

The shear and peel stresses were both assumed to be uniform across the adhesive thickness, and it can also be seen from Fig.2 that the maximum values of peel stress occur at the ends of the overlap (see. Reference [1], p.23 et seq. for a full description of the equations of Goland and Reissner).

Some investigators choose to use the double lap joint, which is effectively two single laps back to back, to eliminate bending and the transverse or peel stresses. In fact, as is shown in Fig.3, this simply modifies the situation, but leaves large tensile peel stresses at one end of the joint owing to the *internal* bending moments.

Various authors (eg. [4], [5]) have improved on the Volkersen and Goland and Reissner theories, but the improvements are largely cosmetic. In particular, it has been assumed that the adhesive layer ends in a square edge as shown in Fig. 4a. But even a rectangular plate with shear loading on its two opposite sides experience high

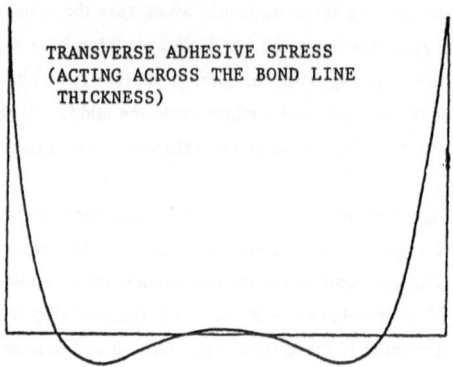

Figure 2 Transverse (Peel) stresses (σ_y) in a single lap joint according to Goland and Reissner.

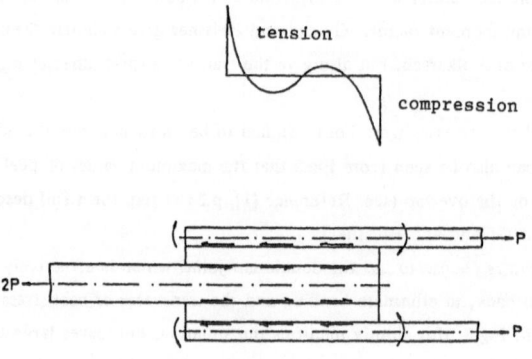

Figure 3.   Bending moments induced in the outer adherends of a double lap joint together with adhesive peel stresses.

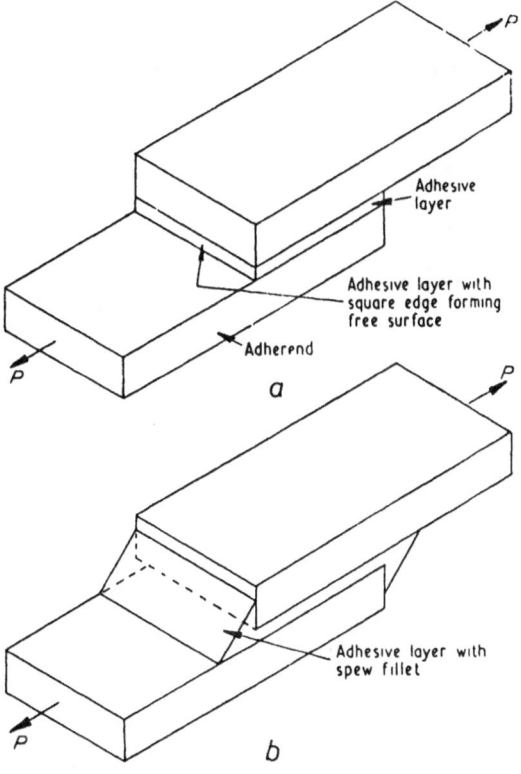

Figure 4. Diagrammatic lap joints to show adhesive layers with a) a square edge,
b) a fillet

tensile and compressive stresses at its corners because of the requirement that the
direct and shear stresses acting on the free surface (edges) must be zero. Thus, in an
adhesive layer with a square edge, similar tensile and compressive stresses must occur
in the corners of this layer because of the free surface. But practical adhesive joints
rarely have a square edge: instead, they are formed with a fillet (see Fig.4b) which is
squeezed out under pressure while the joint is being manufactured. Photoelastic
stress analysis has shown that the position and magnitude of the maximum stress
depends on the edge shape. While the algebraic solutions correctly predict that the
highest stresses are near the ends of the joint, they are unable to take into account
the influence of the fillet. But it is in just these regions of maximum stress where
failure is bound to occur that the assumed boundary conditions of the algebraic
theories are the least representative of reality. Some alternative must therefore be

found, and this is best realised by obtaining a numerical solution to the stress problem which can allow for realistic geometry at the ends of the bondline.

The finite element (FE) method is now a well-established means for mathematically modelling stress (and many other) problems. Its advantage lies in the fact that the stresses in a body of almost any geometrical shape under load can be determined. The method is therefore capable of being used for analysing an adhesive joint with a spew fillet.

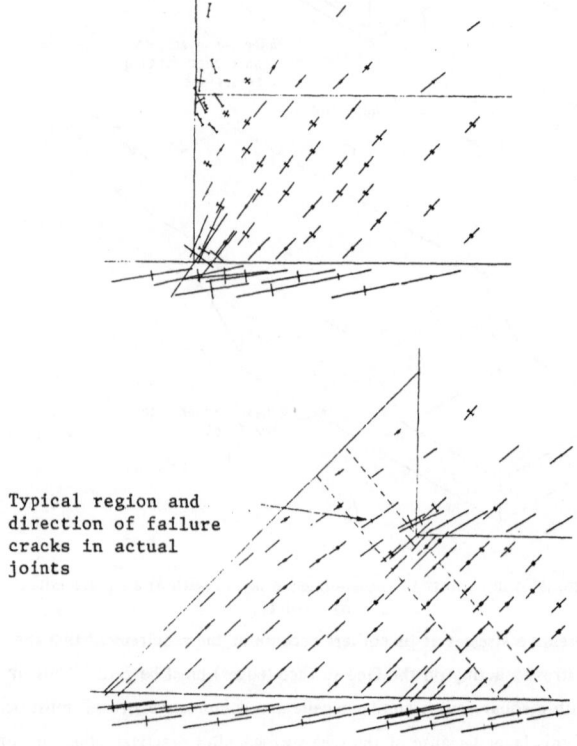

Typical region and direction of failure cracks in actual joints

Figure 5.   (a) Finite element prediction of the principal stress pattern at the end of a square-edged adhesive layer; b) Finite element prediction of the principal stress pattern at the end of an adhesive layer with a 0.5 mm fillet.

Fig.5a shows the stress pattern at the end of a square-edged adhesive layer in a typical aluminium/aluminium lap joint bonded with a structural epoxy adhesive. The highest tensile stress exists at the corner of the adhesive adjacent to the loaded adherend and represents a stress concentration of at least 10 times the average applied shear stress.

The influence of a fillet on the stress pattern is shown in Fig.5b which is at the tension end of a double lap joint. Even though only a very small triangular fillet, 0.5mm high, was used, the stress system is very different from that of Fig.5a. Also, it can be seen that the adhesive at the ends of the adhesive layer and in the spew fillet is essentially subjected to a tensile load at about 45° to the axis of loading. The highest stresses occur near the corner of the unloaded adherend because the 90° corner introduces a stress- concentrating effect. As the maximum stress occurs within the fillet and not at or near the adhesive surface, it is unlikely that the approximation to the spew shape by the triangular fillet has a significant effect on the stress distribution.

Observation of the failure of aluminium to aluminium lap joints bonded with typical structural adhesives shows that cracks are formed approximately at right-angles to the directions of the maximum principal stresses predicted by the elastic finite-element analysis. In general, these cracks run close to the corners of the adherends as shown in Fig.5b. Thus, it can be proposed that failure in a lap joint is initiated by the high tensile stresses in the adhesive at the ends of the joint. Cohesive failure of the adhesive occurs in this manner in normal, well-bonded joints. Under further loading, the initial crack in the fillet is turned to run along (or close to) the adhesive-adherend interface. It meets a similar crack running in the opposite direction.

Inspection of the surfaces of a fractured single or double lap shear test will show the cracks shown schematically in Fig.6.

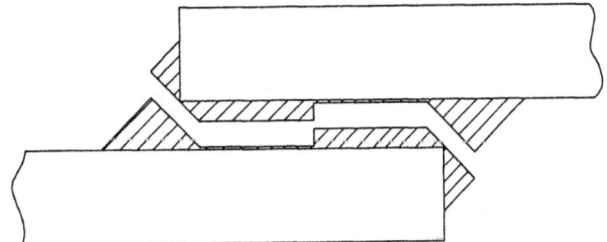

Figure 6. Diagram of failure surfaces of a single-lap joint.

When the adhesive or adherend are loaded beyond yield, the simple theories described so far become inadequate. The only real solution for analysing such behaviour is to use large-displacement, large-rotation elastic-plastic finite elements. A description of such an analysis is beyond the scope of this paper, but it is summarised by Adams and Wake [1].

Adams and Harris [6] investigated the effect of adherend and adhesive yielding. The adhesive maximum principal stress distributions are shown in Fig.7 for a case where the adherend properties correspond to a relatively low strength aluminium alloy (a 0.2% proof stress of 110 MPa) and the adhesive is linearly elastic. Under the action of tension and bending, the adherends begin to yield at an applied load of approximately 1.5kN. At 3 kN, the adherend plastic deformation has had two effects on the adhesive stresses. Firstly, it has led to a reduction in the peak stress

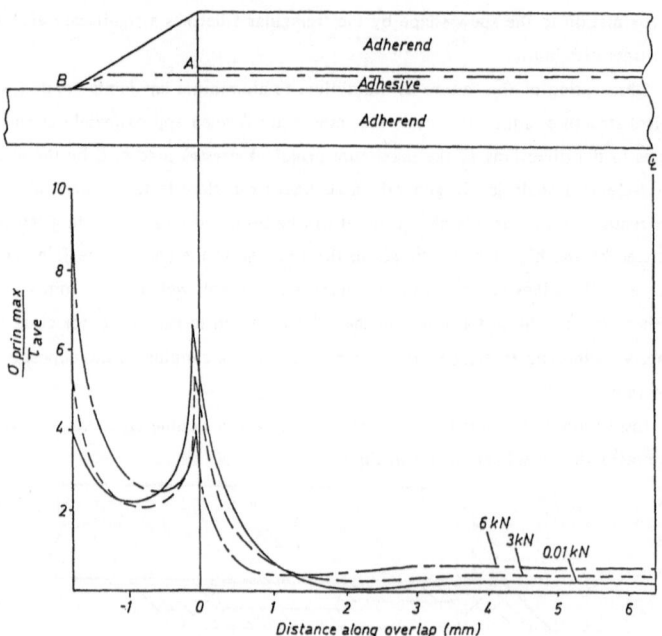

Figure 7. Normalised maximum principal stress distributions along the adhesive layer with adherend yielding, at various applied loads (elastic adhesive)

concentration at the end of the overlap, at point A, over and above that for the elastic case, as a result of the enhancement of the joint rotation. Secondly, the concentration of adhesive stress at the ends of the fillet, point B, has increased, owing to the maximum adherend deformation occurring adjacent to this point producing a localised increase in the differential shear effect. At an applied load of 6 kN, further plastic deformation in the adherends has taken place, the peak at point A is further reduced but, more significantly, the peak at B has dramatically increased, and the stresses at this point are now the highest in the adhesive. It may be concluded, therefore, that when adherend plastic deformation takes place, the joint strength will be reduced and, at the same time, failure will no longer initiate from point A, but from point B. The two types of failure observed in practice, designated as types I and II, are illustrated in Fig.8, type II being indicative of adherend plastic deformation.

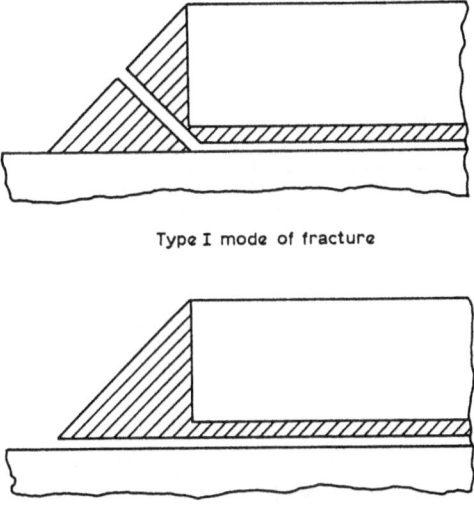

Type I mode of fracture

Type II mode of fracture

Figure 8. Modes of fracture in lap joints with adhesive and adherend yielding.

The other extreme of behaviour is when a high-strength, low ductility adherend is combined with a ductile adhesive. Fig.9 shows the computed principal strains in a

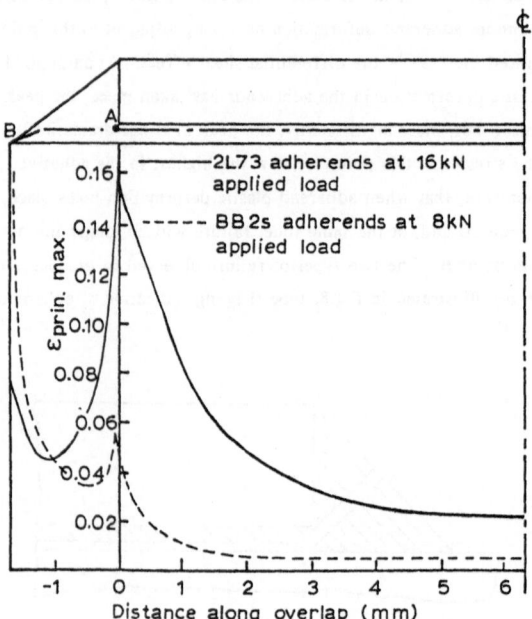

Figure 9. The effect of adherend yielding on the maximum principal adhesive strain distribution for high yield (2L73) and low yield aluminium (BB2S) adherends. The adhesive is a rubber-modified epoxy.

CTBN (carboxyl-terminated butadiene-acrylonitrile) adhesive between two 2L73 high-strength aluminium adherends. This combination gave a joint strength of 16 kN but, when ductile (low yield strength) adherends were used with the same adhesive, the joint strength was reduced by a half to 8 kN.

The important point to be grasped here is that it is essential when doing lap-shear tests that the correct grade of adherend is used. If a low-yield adherend is substituted for a high-yield material, the adhesive will appear to be weaker on the basis of the equation $\tau = P/bl$, which is the normal ASTM 1002 test. See, for example, Fig.49 in Reference [1].

There is a complete range of adhesive behaviour between the two extremes given here, where the failure criterion may not be based on a single parameter, such as maximum principal stress or strain. Further investigation is required into failure criteria for adhesives in order that the limits to the strength of simple lap joints, as well as more complex bonded structures, may be determined under a variety of loads.

While adhesives in a single lap joint appear to be loaded in shear, they do not fail in shear. It is strongly contended that few, if any, structural adhesives fail in shear but that they fail in tension when the principal stresses or strains reach some limiting value.

Various forms of double lap joints are used in adhesive testing. A common form used industrially is to ASTM D 3165-73, as shown in Fig.10. Two sheets are bonded

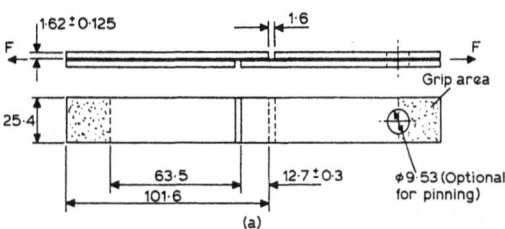

Figure 10. Double-lap joints to ASTM D 3165-73

together and the specimens cut from them. With film adhesives and a press, reasonably uniform bond-line thickness can be achieved. However, liquid or paste adhesives tend to get squeezed out, leading to variable bondlines, and it is difficult with sheets to avoid entrapped air and to vent volatiles. The sheets are cut to form the test area. The main problem is in defining the cut: too deep and the adherends are weakened (especially laminates), too little and the adherend carries some load, and unknown but not insignificant damage is done to the adhesive in the highly-stressed region at the joint ends. The ASTM D 3165-73 specimen is not a true double lap joint but is really a single lap with its adherends stiffened outside the test section.

In the true double lap joint, shown in Fig.3 together with its peel stress distribution, it is difficult to make the two bond lines have the same thickness. The great time and care needed for preparing good specimens is not rewarded by improved results.

Finally, the so-called thick adherend test (ASTM D 3983-81) is a development of the test shown in Fig.10. Differential straining is reduced by using thick adherends, and these are also stiff in bending. Unfortunately, the adhesive is in a far from uniform stress state, as is popularly believed, and there exist substantial transverse peel loads (see Reference [1], p.119, and Renton [7]). In effect, the thick adherend shear test was designed for low modulus adhesives, but it has been used for structural epoxies (and similar high-modulus adhesives) and this gives misleading results. If one is prepared to use the rather fiddly extensometers, this test can be used to obtain accurate shear modulus data -ie. in the elastic region. *But because of the substantial stress (and strain in the plastic case) concentrations at the ends of the overlap, it cannot be used for determining the full elasto-plastic stress-strain curve to failure.*

BUTT JOINTS

Various tests have been developed which purport to give the mechanical properties of adhesives by using butt joints in tension or shear (it is almost impossible to fracture such joints in compression!). ASTM has a variety of tests in this category (see Reference [1] p.295) and the principles of these are briefly reviewed below.

Since some adhesives have a strong exothermic reaction on curing, these can only be made in the thin-film form where the adherends can act as a heat sink. Thus, butt joints provide an apparently convenient, direct and simple means of stressing these adhesives. In the elastic region, a simple, closed form analysis is sufficient, although this is less easy to apply when yield occurs. On the face of it, the stress distribution is simple. However, the problem remains of end effects. If joints are to be loaded to failure and if the failure stress is to mean anything, then it must be the true stress and not a convenient but misleading approximation by taking the stress as the load divided by the area.

ASTM E229-70 specifies a test in which an annular ring of adhesive is sheared between two thin cylinders, leading to the term 'napkin-ring test'. This form of specimen minimises the variations of shear stress in the adhesive and has been used by various authors for measuring the shear strength of adhesives. The shear stress distribution calculated from simple elasticity theory is:

$$\tau_{z\theta} = \frac{2Tr}{\pi(r_o^4 - r_i^4)}$$

where $\tau_{z\theta}$ is the shear stress at a radius r caused by an applied torque T, while r_i and r_o are the inner and outer radii of the annulus. When bonding the specimen, it is usual for some of the adhesive to be squeezed out to form a 'spew fillet' and this may be expected to modify the simple, theoretical, stress distribution. Over much of the joint, the fillet reduces the stresses slightly. Unfortunately, at the outer and inner circumferences, adjacent to the substrate, there is a large stress concentration [8]. Removing the fillet minimises the stress concentration and this is recommended in the ASTM E229-70 test, albeit for other reasons. Adams and Coppendale [9] estimated that the fillet gave a stress increase of about 1.8 times at the edge, compared with the situation with no fillet. It is therefore important with napkin ring (and similar shear tests) to remove the fillet before testing: otherwise, misleading strength results will be given.

Adams and Coppendale [9] recommend the use of a solid (ie. not annular) torsional butt joint. This is easier to form than the napkin ring arrangement and only the outer fillet need be removed, which is fairly easy. A simple graphical correction (see Fig. 85 in Reference [1]) of the torque-twist curve can be used to obtain true shear stress-strain data, even when the adhesive yields.

Butt joints for testing adhesives in tension are also usually designed with a circular cross-section to facilitate manufacture and to maintain symmetry. In this case, it would appear to make little difference whether they are annular or solid. In a butt joint subjected to a tensile load, the adhesive is restrained in the radial and circumferential directions by the adherends. In the absence of this restraint, the adhesive would tend to contract radially with respect to the adherends because of its much lower modulus. The presence of the adherends has the effect of inducing radial and circumferential stresses in the adhesive, so increasing the stiffness of the joint. The simplest analysis makes the assumption that the radial and circumferential strains in the adherend and the adhesive are zero, in which case the radial and circumferential stresses are given by:

$$\sigma_r = \sigma_\theta = \left(\frac{\nu}{1-\nu}\right)\sigma_z$$

where ν is Poisson's ratio of the adhesive and σ_z is the nominal applied axial stress. The apparent Young's modulus E^1 (defined as the applied axial stress divided by the axial strain) is given by:

$$E^1 = \frac{\sigma_2}{\epsilon_2} = \frac{E\,(1-\nu)}{(1+\nu)(1-2\nu)}$$

where E is Young's modulus of the adhesive. Taking account of the adherend strains by assuming that the radial strain in the adhesive is equal to the Poisson's ratio strain in the adherends, we have,

$$\epsilon_r = \epsilon_\theta = -\left(\frac{\nu_a}{E_a}\right)\sigma_z$$

where ν_a and E_a are the Poisson's ratio and Young's modulus of the adherends. The radial and circumferential stresses now become:

$$\sigma_r = \sigma_\theta = \left(\nu - \frac{E\nu_a}{E_a}\right)\left(\frac{\sigma_z}{1-\nu}\right)$$

These simple analyses ignore the requirement that the radial stress must be zero at the free boundary of the adhesive, which implies that shear stresses must exist in the adhesive layer.

Using the same finite element techniques as for torsion, Adams, Coppendale and Peppiat [8] analysed butt joints in tension to elucidate the various radial, circumferential and axial stress distributions. Major stress concentrations were found near the outer edges, the axial stress increase being estimated to be about 2.5 times the average: it is also in a region of high, complex stresses. The presence of a fillet tends to reduce this stress concentration, in contrast to the case in shear discussed above.

Further work by Adams and Coppendale [9] considered what happens when the adhesive yields. They used a pressure-dependent yield criterion (ie. one in which the von Mises deviatoric stresss causing yield increases linearly with hydrostatic pressure). They postulated that, under compressive loads, the adhesive in a butt joint should not yield, but if yielding does begin to occur then, since Poisson's ratio will increase towards 0.5, the yield zone would be very restricted. Tensile butt joints *appear* to provide a simple means of establishing the direct stress-strain and tensile strength properties of an adhesive. However, it is difficult to measure the small deflections involved (compensating also for adherend strains (Reference [1] p.122)) and the large stress concentrations at the periphery of the joint imply that any strength values so obtained are likely to be meaningless. Other problems, such as avoiding bending, only compound the difficulties. *It is therefore strongly recommended that axial butt joint tests for adhesive properties be avoided.*

PEEL TESTS

Various forms of peel test are used to assess the performance of structural adhesives. In effect, this form of test deliberately stresses the adhesive in a very small region, subjecting it to a large tensile stress, although there is usually a complex stress situation present. The stress situation is not easily assessed and peel is normally used to *compare* adhesives rather than to measure their properties.

The basis of the peel test is shown schematically in Fig.11. A thin substrate, say aluminium alloy, is bonded to a thick substrate which is itself clamped (or bonded) to a rigid support. A force P is applied at some angle θ and the strip is 'peeled' off. In testing, care has to be taken to keep the angle θ constant: this is not easy.

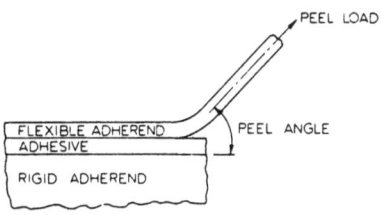

Figure 11. Diagrammatic representation of the peel test

Crocombe and Adams [10] analysed the test, allowing for both the adhesive and the strip to become plastic. They showed that the main stresses in the adhesive were such as to drive a crack towards the interface with the flexible adherend. The actual transverse stress at the point where fracture was expected to occur was a function of the peel angle.

Figure 12 shows various forms of the peel test. The most commonly used is the T peel (effectively a 90^0 peel test) as specified by ASTM D 1876-72. It has the advantage that the two ends can be gripped in the jaws of a testing machine and pulled without complications, other than ensuring that the unfractured portion remains at 90^0 to the pulling direction. The results are usually expressed as load divided by the strip width (1b/in or N/mm).

The climbing drum test (Fig.12) is used for peeling the face plates from honeycomb sandwich constructions and is specified in ASTM D 1781-76. The floating roller arrangement shown in Fig.12 is described in ASTM D 3167-76. In theory, the arc at which the skin is peeled away in both the climbing drum and the floating roller tests is controlled by the roller geometry, but in practice the conformity is not so certain.

On the whole, the floating roller gives the most repeatable results, while the climbing drum is preferred for honeycomb, and the T-peel is the easiest to carry out.

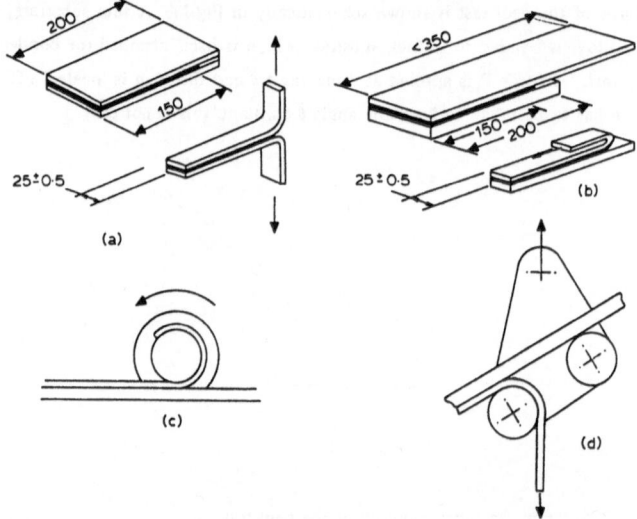

Figure 12. Various forms of the peel test: (a) 180^0 T-peel test for flexible-to-flexible assembly; (b) 180^0 peel test for flexible-to-rigid assembly; (c) climbing drum test; (d) floating roller test (dimensions in mm).

Before leaving the consideration of peel tests, it is necessary to look at the test often referred to as the 'Boeing wedge test' and now specified in ASTM D 3762-79 and shown in Fig.13. The stresses are generated by forcing the wedge between the two bonded plates, thus forming peel (cleavage) stresses across the bond line. This test is not normally used for quantitative or short term testing, but is a cheap and sensitive way of investigating environmental effects and differences in surface preparation. The test yields information on crack growth rate and as to whether the fracture is between the adhesive and the substrate, or wholly within the adhesive.

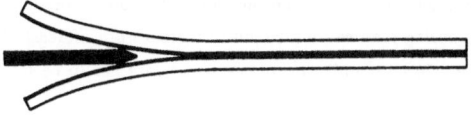

Figure 13. Boeing wedge test

CONCLUSIONS

A brief account has been given above of some of the common test methods used for assessing the performance of adhesives. These tests fall into two basic categories, production and scientific.

Tests such as the lap-shear test in all its forms are essentially comparative. Although numerical values may be given for lap-shear strength, these do not in any way represent the true shear strength of the adhesive. As long as the testers do not try to interpret the results as being scientifically exact, then a healthy situation exists.

The peel test is similarly a comparator. At one time, researchers even believed that the area under the load-extension curve in peel equalled the work done in separating the surfaces. Unfortunately, they failed to account for the substantial element of plastic work in the flexible adherend.

The thick-adherend shear test can only be used for giving adhesive shear moduli. Under no circumstances should its results be relied on for indicating the shear strength (or strain) of an adhesive. It has no advantage over the napkin ring or solid butt joint in torsion. The tensile butt joint test is as misleading as the thick adherend shear test. Apparently simple to interpret, it is difficult to carry out accurately and has a far from simple stress distribution.

If researchers and designers need true strength and modulus values, great care must be taken in carrying out the tests and interpreting the results, since most of the tests involving adhesive joints do not give directly interpretable values. It is suggested that the best test for determining the mechanical properties of an adhesive is to make specimens from bulk samples, taking care that these have reached the same state of cure as the investigator believes to be the case. For some adhesives, exothermic reactions on curing may cause a runaway temperature increase, leading to over-heating and damage. In most bulk specimens, great care has to be taken to avoid air bubbles.

REFERENCES

1.   Adams, R.D. and Wake, W.C., 'Structural adhesive joints in engineering'. Applied Science Publishers, London and New York, 1986.

2.   Volkersen, O., 'Die Nietkraftverteilung in Zugbeanspruchten mit Konstanten Laschenquerschritten', Luftfahrtforschung, 1983, Vol.15, pp.41-47.

3.   Goland, M., and Reissner, E., 'Stresses in cemented joints', Journal of Applied Mechanics, Transactions of the American Society of Mechanical Engineers, 1944, Vol.66, pp.A17-A27

4.   Renton, W.J. and Vinson, J.R., 'The efficient design of adhesive bonded joints', Journal of Adhesion, 1975, Vol.7, pp.175-193.

5.   Allman, D.J. 'A theory for the elastic stresses in adhesive-bonded lap joints', Quarterly Journal of Mechanics and Applied Mathematics,1977, Vol.30, pp.415-436.

6.   Harris, J.A. and Adams, R.D., 'Strength prediction of bonded single lap joints by non-linear finite element methods', International Journal of Adhesion and Adhesives, 1984, Vol.4, pp.65-78.

7.   Renton, W.J., 'The symmetric lap-shear test - what good is it?' Experimental Mechanics, 1976. Vol.33, pp.409-415.

8.   Adams, R.D., Coppendale, J., and Peppiatt, N.A., 'Stress analysis of axisymetric butt joints loaded in torsion and tension', J. Strain Analysis, 1978, Vol.13, pp.1-10.

9.   Adams, R.D. and Coppendale, J., 'The elastic moduli of structural adhesives', in Adhesion 1, ed. K.W. Allen, Applied Science, London, 1977, pp.1-17.

10.   Crocombe, A.D. and Adams, R.D. 'Peel analysis using the finite element method', J. Adhesion, 1981, Vol.12, pp.127-139.

2

AN EXPERIMENTAL ASSESSMENT OF THE COIN-TAP TECHNIQUE FOR DETECTING DEFECTS IN ADHESIVELY BONDED SHEET STEEL JOINTS

P A Fay and D F Elms

Ford Motor Company Limited
Research and Engineering Centre
Laindon, Basildon, Essex, SS15 6EE

INTRODUCTION

It is likely that most bonded joints contain some imperfections or defects. The significance of these will depend not only on their extent, but also on their position, the loading and environmental conditions which the bond will experience in service and, fortunately, in many cases, their existence will not be detrimental. It has been shown, for instance, that a debond extending to over 50% of the area in the centre of a simple overlap has very little effect on the initial strength of the joint [1].

However, as adhesives become more widely used in critical structural applications, such as aerospace and automotive structures, a parallel need emerges for methods for detecting critical defects during the assembly and the service life of such structures.

A large number of NDT techniques are currently available for inspecting bonded joints [2,3,4]. Unfortunately, amongst these, there is no technique which, when applied to a bonded joint, will non-destructively give a absolute measure of its strength. Some techniques attempt to correlate strength with some other measurable property such as bond area, stiffness or damping and, whilst these measures do not necessarily give direct correlation with joint strength, they are still used in certain applications for indicating the presence of defective joints.

However, most of these techniques have limitations when applied to bonds between contoured sheet steel components with varying bondline thickness, such as those found in automotive structures. Furthermore, these methods are often impractical for mass production since they are time consuming (both to carry out and to interpret) and they usually require unsuitable processes, such as complete immersion.

A much simpler, alternative NDT technique based on the age old coin-tap overcomes some of these limitations and it has already been used successfully for inspecting bonded honeycomb structures. The object of this investigation was to determine the capability of a commercially available instrument based on the coin-tap technique in detecting various defects and geometric characteristics in adhesively bonded sheet steel joints in order to assess its potential as an NDT technique for bonded automotive structures.

PRINCIPLE OF THE THE COIN-TAP TECHNIQUE

The coin-tap technique is based on the simple observation that a "defective" area of a bonded component when tapped, sounds "dead" compared with a "good" area. This is different to the apparently similar "wheel-tap" technique where the whole component sounds "dead" when tapped if a defect is present, compared with the sustained, clear note produced by a good component. In essence, the wheel-tap test is a global test identifying defective components without actually identifying the location of the defect, whereas the coin-tap technique gives an assessment of integrity at each test site, to which some criterion has to be applied to judge the overall quality of the component.

Whilst the coin-tap technique is based on a very simple phenomenon, it has been shown to be quite effective, particularly on honeycomb structures. Segal and Rose [2] commented that, bearing in mind the low cost of equipment (25 cents compared with over 100,000 dollars for a large immersion ultrasonic system), in many cases the coin-tap technique may prove to be a better option than the more sophisticated (and much more expensive) techniques which are only slightly more effective.

MECHANICS OF THE THE COIN-TAP TECHNIQUE

The mechanics of the coin-tap technique have been investigated further by Adams and Cawley [5,6] who have shown that, in addition to audible changes, the form of the force input to a structure during a tap is also changed by the presence of defects and they have demonstrated that this can be used as the basis of a reliable NDT method, eliminating the subjectivity of the human ear in assessing the frequency content and amplitude of the sound produced in the simple coin-tap test.

THE TAPOMETER

This led to the development of an instrument called the Tapometer which uses an electro-magnetic relay to accelerate a tapping head towards a structure. A piezoelectric force transducer incorporated in the hammer provides a force-time trace for the impact. The trace from a test tap can be compared directly with a reference trace from a known good test site. In addition, force-frequency spectra can be produced, the shapes which can, in turn, be used to provide a quantitative measure of integrity.

Further development of the instrument was carried out by by Rolls Royce MatEval Limited, which led to the production of a commercially available portable instrument [7] (which now unfortunately costs more than the 25 cents quoted by Segal and Rose [2]). The instrument was primarily designed for inspecting composites in the aerospace industry and, in this field, it has proved successful in locating defects such as delaminations lying parallel to the inspection surface.

The MatEval instrument was used in this investigation. It consists of two major components, the tapping head and the control unit (see Figure 1).

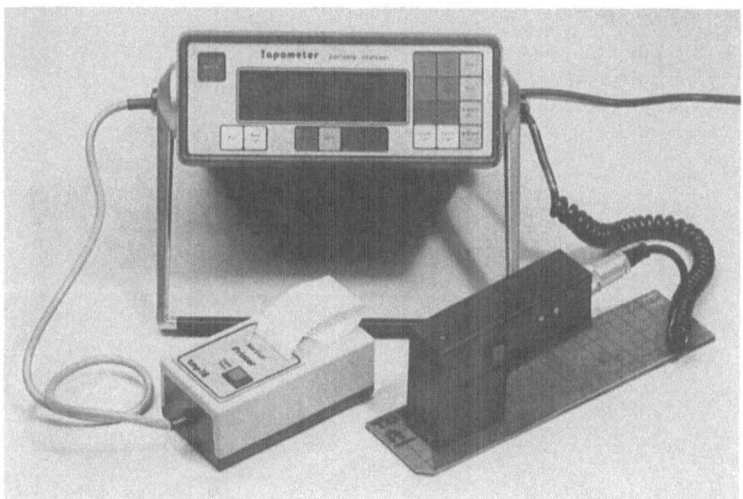

Figure 1. The Tapometer Instrument

The head incorporates a solenoid hammer and an piezoelectric accelerometer to measure the response. The control unit is microprocessor driven with dedicated, menu-driven software to guide the operator through the set up and operating procedures. A dot matrix LCD screen displays the results of an individual test tap and a reference tap on a known good region. The display can be in the form of either force-time traces or force-frequency spectra. The Tapometer compares these traces and calculates a numeric value called the "difference value" (see below) which gives an indication of bond integrity. Audible and visual alarms can be activated if this value falls below a user defined threshold. The unit is also capable of printing a tabular output of the data.

DIFFERENCE VALUE CALCULATION

Typical force-time traces from good and defective areas of an adhesively bonded structure are shown in Figure 2. The effect of the defect is to make the impact less intense and of longer duration than that on a good area. Either of these differences in characteristic could be used as a criterion for assessing good and defective areas but, as Adams and Cawley have pointed out [6], the absolute value of the peak force will depend on the magnitude of the applied impulse and, although the time period is less dependent on the magnitude of the impulse, it is not easy to measure accurately.

A better approach is to determine the frequency content of the traces using Fourier transforms and compare the corresponding force-frequency spectra. This technique allows for more accurate discrimination between the impulses. Typical spectra from good and defective areas are shown in

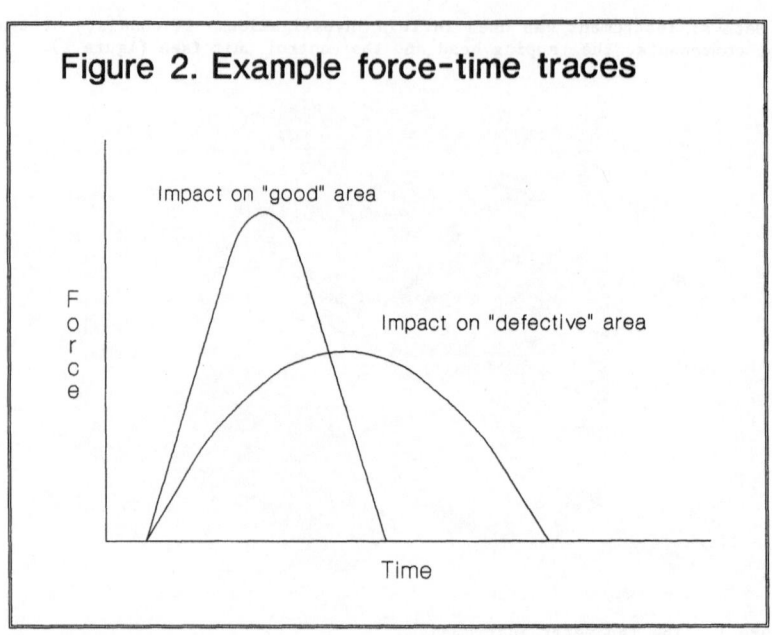

Figure 2. Example force-time traces

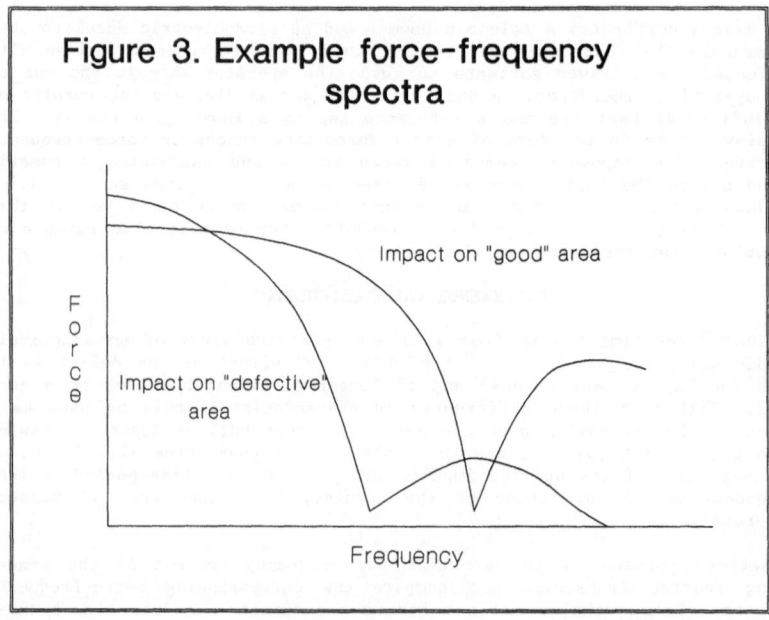

Figure 3. Example force-frequency spectra

23

Figure 3. The impulse on the defective area has more energy at low frequencies but falls quickly with increasing frequency whereas the impulse on the good area demonstrates a much lower rate of decrease of energy with frequency. Adams and Cawley [6] found that the shape of these traces varied little with changes in impulse and felt that identifiable differences between the shapes of good and defective spectra could be used as a sensitive pass/fail criterion.

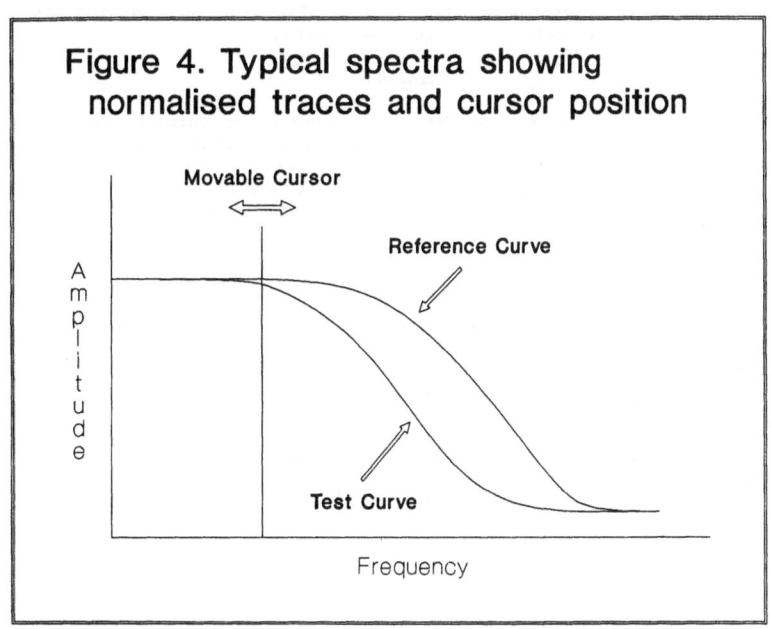

Figure 4. Typical spectra showing normalised traces and cursor position

The Tapometer carries out a comparison of the spectra by first normalising the test and reference traces. In most cases it does this by setting the curves to the same magnitude at the lowest frequency (See Figure 4). A cursor is then placed at a selected frequency on the spectrum and the difference value is calculated as follows:

$$\text{Difference Value} = 100 - \frac{100(A/B - C/D)}{A/B}$$

Where:
A = Area under the reference curve to the right of the cursor
B = Total area under the reference curve
C = Area under the test curve to the right of the cursor
D = Total area under the test curve.

This gives a value of 100 for the "good" reference curve and lower figures for spectra produced by tapping on defective areas. These figures are sometimes quoted as percentages but this is extremely misleading because tapping an area where there is no bond at all (eg where there is no adhesive!) does not produce a value of zero and no correlation exists between the difference value and values of measured joint strength. In this paper, the authors use the term "difference value" to refer to the figures produced from the above equation and make no reference to percentages to avoid confusion.

The difference values obviously depend on the position of the cursor. This is set by the operator and the factors affecting this setting and the other calibration parameters are discussed below.

CALIBRATION

Like most NDT instruments, the Tapometer requires calibration before use. There are two stages to this. Firstly the instrument must be adjusted so that it records and analyses the traces correctly. Once this is done, a reference curve is recorded by tapping on a known, "good" part of the component and all subsequent test traces are compared with this reference curve.

Calibration of the instrument is carried out either by keyboard inputs for the necessary parameters (gain and bandwidth) or by using the in-built "self learning" auto-calibration facility. Once these values have been set, it is then necessary to set the position of the cursor so that the relevant areas of the traces are compared. Eight sets of calibration values can be stored in the Tapometer's internal memory.

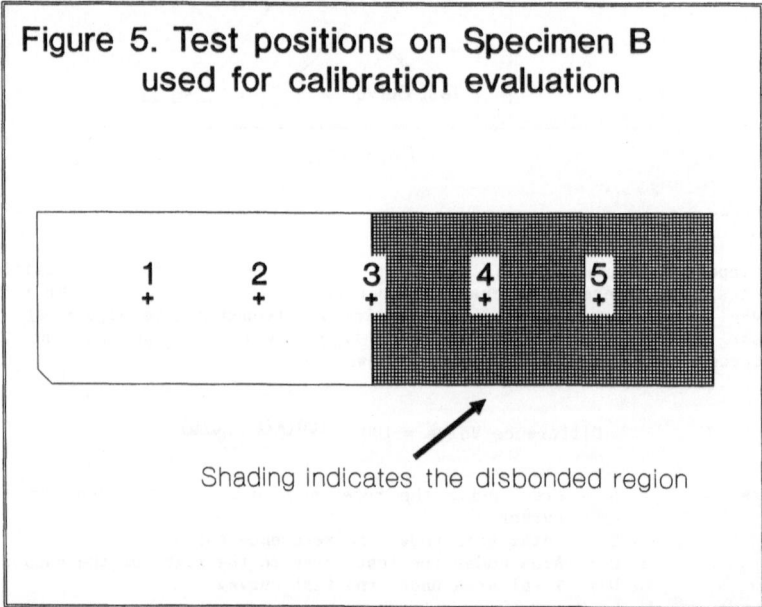

Figure 5. Test positions on Specimen B
used for calibration evaluation

1 2 3 4 5

Shading indicates the disbonded region

As part of this investigation, a series of evaluations were carried out on the initial setting up of the instrument to determine the effect of varying the values of gain and bandwidth and the position of the cursor.

The specimen selected for these tests contained a clearly defined boundary between areas of good and defective bond. Five test positions were selected along the specimen going from the good area to the disbonded region as shown in Figure 5.

Gain
Gain was the first parameter investigated. The chosen value affects the amplification of the results. If the setting is too low then sensitivity is lost and if too high then the test data overshoots the display screen and is distorted. The effect of testing at the five points along the calibration specimen with different gain settings can be seen in Figure 6. The transition between good bond and disbond occurs at about test position 3. It can be seen the 24 dB value gives a clear indication of the transition and gives considerably reduced difference values for the disbonded region. Other values of gain show a less clear transition. From this graph there is also an indication that a difference value of 80 is a reasonable criteria for discriminating between good & bad bond areas.

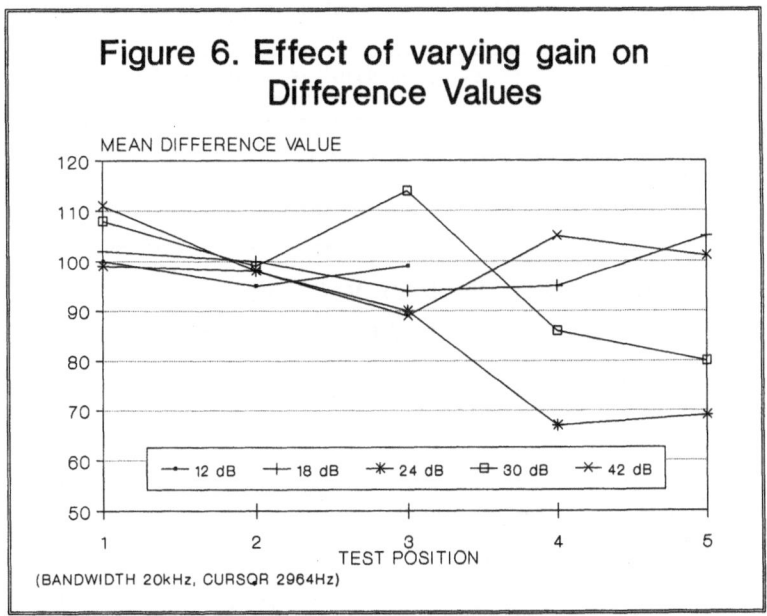

Figure 6. Effect of varying gain on Difference Values

Bandwidth
The bandwidth is the range of frequencies over which the spectra are calculated and analysed. Its value is chosen so that the most relevant section of the frequency response curves are analysed. If the setting is too high then a large area of the high frequency end of the trace is analysed which is identical on both the reference and the test traces, thus reducing the difference which can be detected between the two curves. On the other hand, if it is too low, then the results are distorted because the area where the curves diverge is not included. In Figure 7 you can see that as the bandwidth was increased during tests on

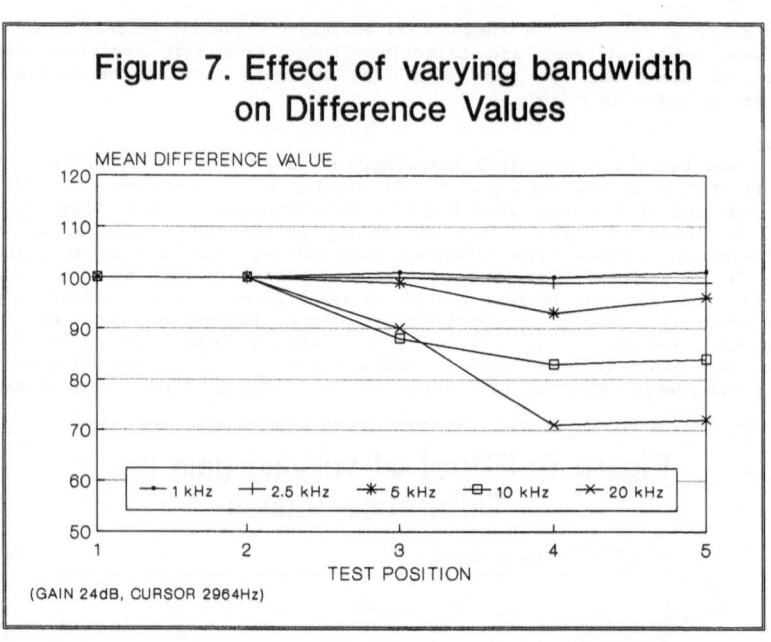

Figure 7. Effect of varying bandwidth on Difference Values

(GAIN 24dB, CURSOR 2964Hz)

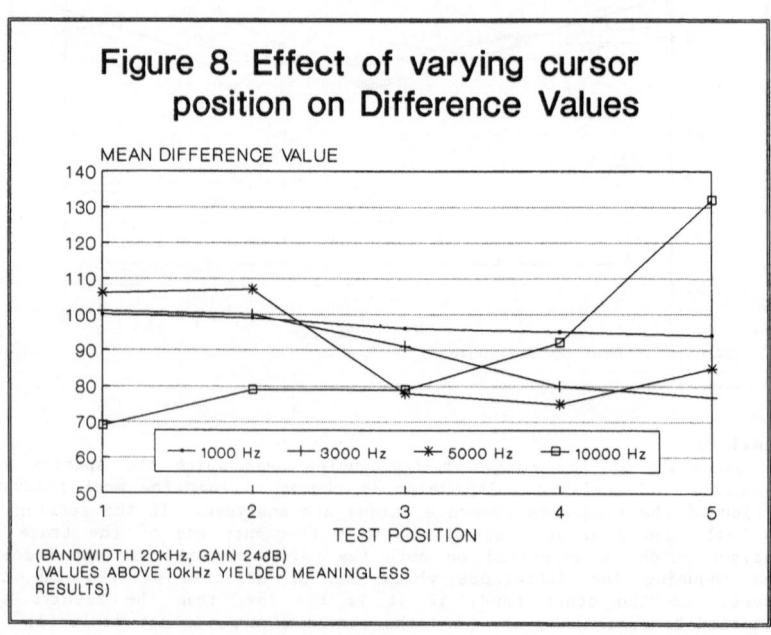

Figure 8. Effect of varying cursor position on Difference Values

(BANDWIDTH 20kHz, GAIN 24dB)
(VALUES ABOVE 10kHz YIELDED MEANINGLESS RESULTS)

the calibration specimen, the detection of the disbonded section became more evident as more of the full frequency curve is analysed and a value of 20 kHz was found to give a reasonable change in difference value between the good and defective areas.

Cursor Position
The position of the cursor does not affect the frequency spectra generated during a test. It only affects the comparison of the test spectrum and the reference spectrum. As the cursor is moved from right to left the sensitivity increases and it reaches an optimum setting when it is the region where the calibration curve and the test curve start to diverge. In Figure 8 it can be seen that too low a setting reduces the detection ability between good and defective areas and too high a setting gives results which are meaningless. A cursor position in the region of 3000 Hz generates results which again show a clear transition.

This evaluation resulted in the following values being selected for these parameters to give the best sensitivity on the steel joints used in the investigation:

Gain	24 dB
Bandwidth	20 kHz
Cursor Position	2964 Hz

Auto-Calibration
The Tapometer also has a self learning facility which adjusts values of gain and bandwidth to suit a particular test structure. This facility greatly speeds up the calibration procedure on a new, unknown structure and, when tried on the same specimen used in the manual exercise described above, yielded the same values of gain and bandwidth.

Mis-Calibration
In order to calibrate the Tapometer it is necessary to have a section of structure which is known to be sound. As part of this investigation, the effect of calibrating on a defective area by mistake was investigated. An auto-calibration was carried out on the disbonded section of the specimen and a test on this section was saved as the reference curve for subsequent tests. Tests on other parts of the disbonded section produced difference values in the region of 100 but with unusually high scatter. However, when tests were carried out on the good section of this specimen difference values of over 130 were registered, again associated with high scatter. This shows that the Tapometer, after being calibrated on a defective area, is still capable of detecting a "better" area. The experienced operator would be able to detect poor calibration of this kind and could recalibrate the Tapometer if readings in excess of 100 are persistently registered, especially if they are associated with high scatter in the results.

EXPERIMENTAL PROCEDURE

Test Specimens
The Tapometer was then evaluated on a series of test specimens containing a range of different simulated defects and geometric characteristics of joints. The specimens were selected to cover:

a) Typical defects which could occur under production conditions (such as concentrations of process oil and areas of insufficient adhesive).

b) Academic defects which, although unlikely to occur in practice, were used to produce defects of known size, location and severity (such as the inclusion of polypropylene discs in the bondline) in order to assess the sensitivity of the Tapometer.

c) Particular joint characteristics (eg tapered bondlines, curved substrates, varying substrate thickness) aimed to evaluate the Tapometer's ability to accommodate the geometric variations found on automotive structures.

The test specimens were made from rectangular mild steel sheets 76 mm x 305 mm. See Figure 9 for details of the specimen configuration. These dimensions are not typical of joints found on automotive structures and they were chosen for this investigation to suit the shape and size of the Tapometer head.

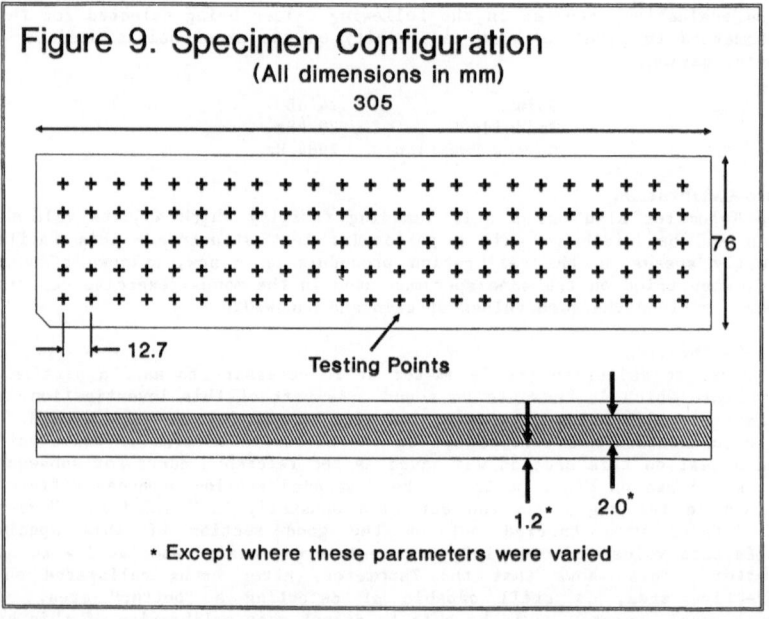

Figure 9. Specimen Configuration
(All dimensions in mm)

The majority of the specimens used flat, 1.2 mm thick sheets and 2.0 mm thick bondlines. In addition, a number of specimens with tapered bondlines (0.5 to 3.0 mm), thicker substrates (2.0 mm) and curved substrates were also included.

A toughened epoxy paste adhesive was used throughout and, because it was a metal filled material, the exact location of most of the "defects" could be determined using X-rays. On the specimens containing disbonded areas, their extent was determined by a visual examination of the specimens.

Further details of the configuration of the individual specimens and the defects incorporated in them are given in the Discussion section below.

TESTING

Once correctly calibrated, the Tapometer was used to evaluate each test specimen. 5 repeat test taps were carried out at each of 115 discrete points as shown in Figure 9 and a mean difference value was calculated for each point. Figure 10 shows the Tapometer being used on one of the curved specimens. The measured difference values are displayed in increments of 3 units only and the typical range of results of repeated tests on a good area was approximately 94 to 100. Scatter on a defective areas was greater, typically ±6 units.

Figure 10. The Tapometer being used to test one of the curved specimens

RESULTS

The results are presented in Figures 11 to 20. The results of the Tapometer tests are presented as graduated shaded plots with the darkest regions indicating test points producing the highest difference readings (ie good areas) and the lightest region the lowest values (ie defective areas).

DISCUSSION

For the purposes of this discussion the specimens were grouped into five categories: those with disbonds, voids, inclusions, geometric characteristics and, finally, a damaged joint.

Disbonds
A number of specimens were prepared with large disbonded regions. Generally, the Tapometer was able to map out the extent of the disbonded region as the following results on specimen A show. This specimen had a complete disbond (produced using large quantities of process oil on the

substrate surface) starting from its right hand end covering almost half of the specimen area. The Tapometer results show visually a clear distinction between the good and defective areas (Figure 11), although there are a number of high values recorded in the disbonded area and vice-versa (particularly around the edges).

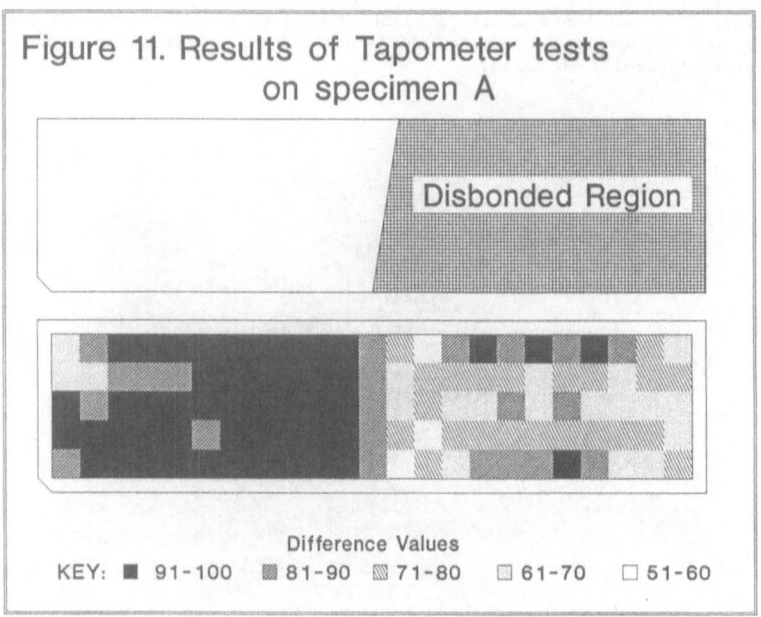

Figure 11. Results of Tapometer tests on specimen A

Disbonded Region

Difference Values
KEY: ■ 91-100 ▓ 81-90 ▒ 71-80 ☐ 61-70 ☐ 51-60

If the difference value of 80 suggested earlier is selected as a pass/fail criterion then an even clearer indication of the disbonded area can be if the results are plotted on a pass/fail plot (Figure 12).

Similar results are seen with two of the other disbonded specimens: B (with a disbond produced using a release agent) and J (with curved substrates) as shown in Figure 12. However on the specimen with the thicker substrates (M), no evidence of the disbond could be seen.

All these results were obtained by tapping on the top face of the specimen which was the side of the joint closest to the disbond. It was found that the extent of the disbonded areas could also be determined by tapping on the lower face, although the individual difference values at each test site were different from those produced from the top face.

Voids
The Tapometer was found to be reasonably effective in detecting large voids (typically greater than 10 mm in diameter), although greater detection would probably be possible if a finer tapping grid was used. As an example, the results of the tests on Specimen I are given in Figure 13 The voided area of the specimen is shown by shading and the plot of the Tapometer results shows that the large void in the centre of the joint is

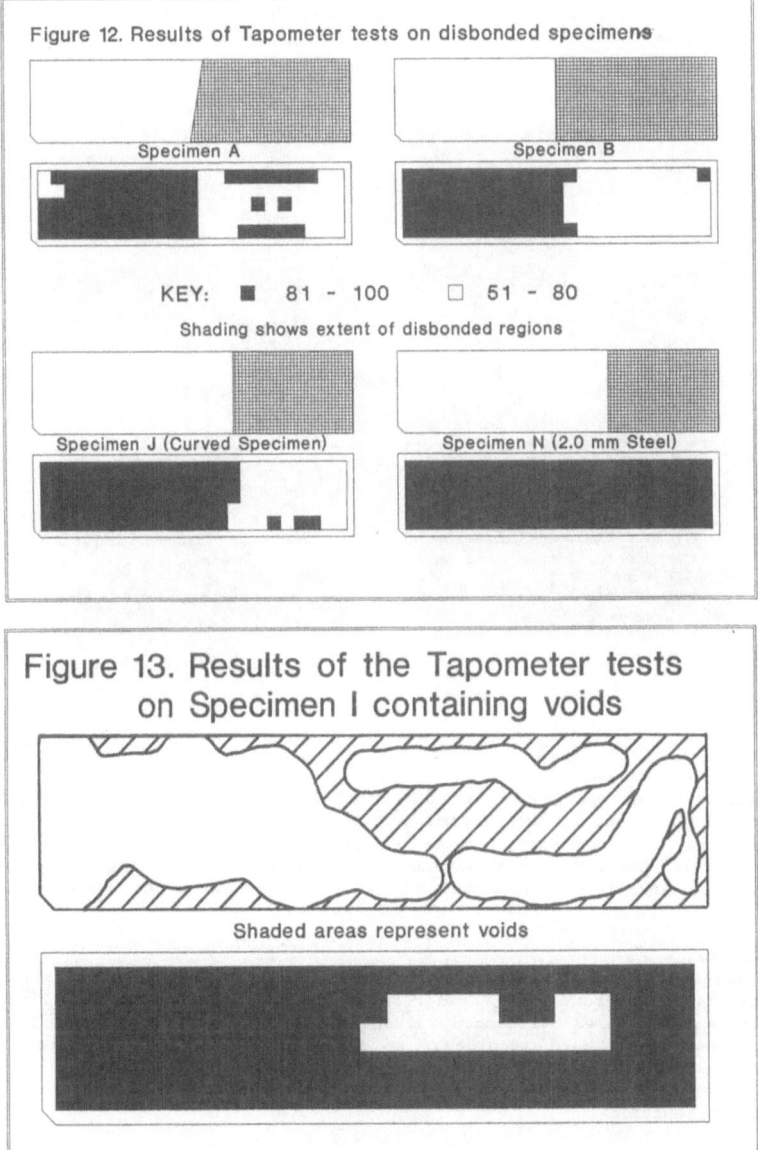

Figure 12. Results of Tapometer tests on disbonded specimens

Specimen A

Specimen B

KEY: ■ 81 - 100   □ 51 - 80

Shading shows extent of disbonded regions

Specimen J (Curved Specimen)

Specimen N (2.0 mm Steel)

Figure 13. Results of the Tapometer tests on Specimen I containing voids

Shaded areas represent voids

KEY: ■ 81 - 100   □ 51 - 80

clearly detectable, although its precise mapping is obviously limited by the spacing of the test taps.

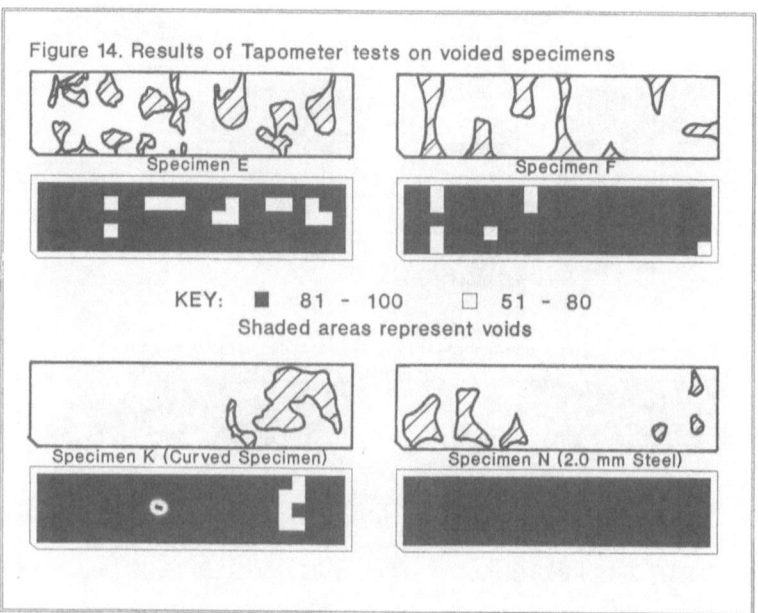

Figure 14. Results of Tapometer tests on voided specimens

Specimen E

Specimen F

KEY: ■ 81 - 100   □ 51 - 80
Shaded areas represent voids

Specimen K (Curved Specimen)

Specimen N (2.0 mm Steel)

The other voided specimens with different levels of voiding or porosity (specimens E, F and K) showed similar results, with most of the the large voids being approximately located on both flat and curved substrates (Figure 14). However, the technique was again found to be less sensitive on thicker substrates (specimen N), where no evidence of voiding was detected.

Inclusions
Two specimens (C & D) were prepared with quite high concentrations of inclusions in the bondlines. The results (Figure 15) show no significant reduction in measured difference values. Even in those areas where lower difference values were recorded, no correlation with the position of actual inclusions was observed.

Geometric Features

Tapered Bondline: Specimen G was prepared with the bondline tapering from 3 mm thick down to zero along the length of the specimen. No other defects were introduced and the entire bond area is believed to be sound. A reference trace was recorded at an average bondline thickness and then the entire area was tested. As the results in Figure 16 show, the vast majority of the specimen was found to be sound.

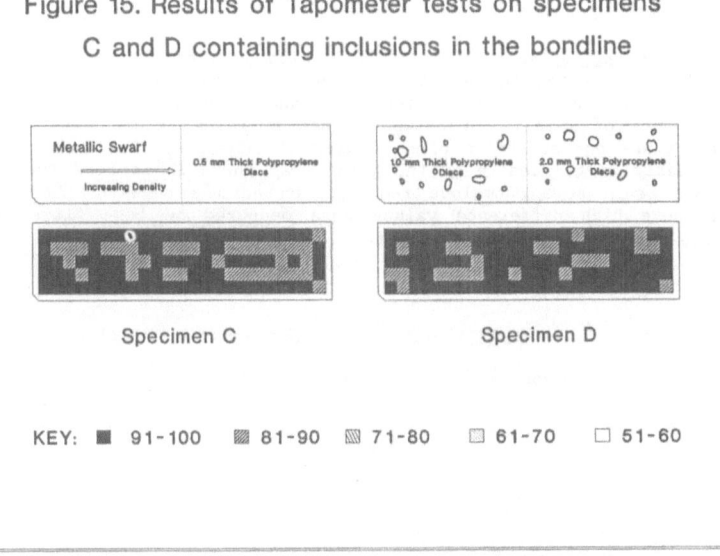

Figure 15. Results of Tapometer tests on specimens
C and D containing inclusions in the bondline

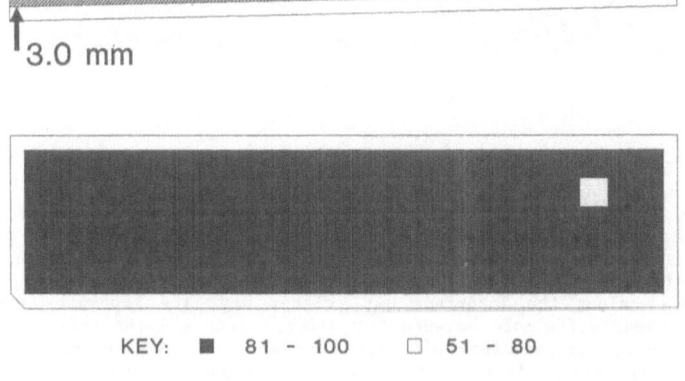

Figure 16. Results of Tapometer tests on
Specimen G with tapered bondline

This is particularly important for applications with varying bondline thicknesses (such as automotive structures) since it implies that the Tapometer can cope with a wide range of bondline thicknesses without the need to recalibrate.

Varying Cross-Section: Specimen L with varying cross-section, see Figure 17, was tested to see if the Tapometer could detect areas where there was a bond underneath the test site and areas where there was just one thickness of metal.   In between these extremes, there were regions were there was one substrate plus a layer of adhesive.   The results show quite clearly in the centre of the joint that there is no bond there but there are again some anomalous readings around the edge of the centre region where high difference values were measured on test sites where there was only one thickness of metal.

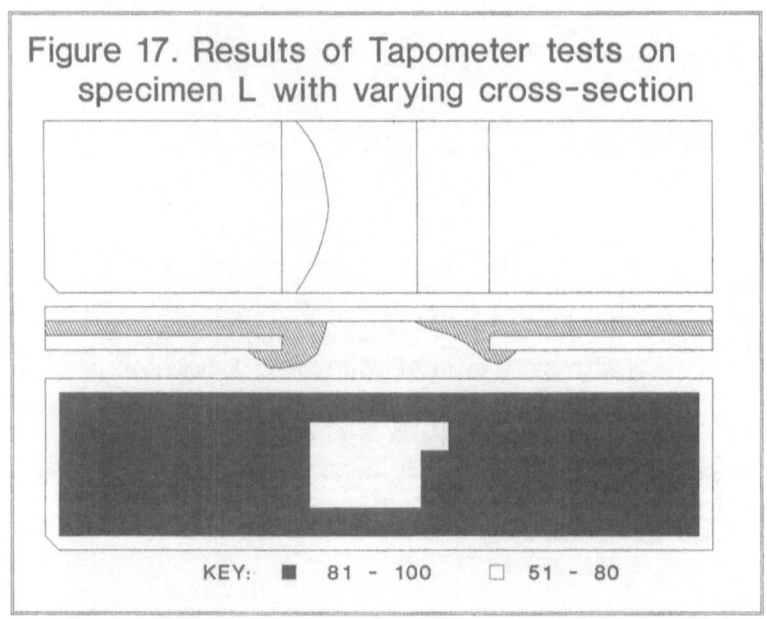

Figure 17. Results of Tapometer tests on specimen L with varying cross-section

KEY: ■ 81 - 100   □ 51 - 80

Damage

The final specimen in this investigation (specimen H) was originally sound and, after curing, a crack was forced down the specimen from the left hand edge in an attempt to reproduce the type of damage which may occur in a vehicle impact.   It is not known whether this crack goes through the adhesive or along the interface but, either way, the Tapometer results showed a clear difference between the damaged area and the area which is still bonded.   The pass/fail plot for this specimen (Figure 18) shows an easily interpretable visual indication of the transition between the good and damaged areas.   It is possible to imagine, therefore, that a simplified instrument could be developed, based on this technique, for use in repair workshops for determining the extent of impact damage.

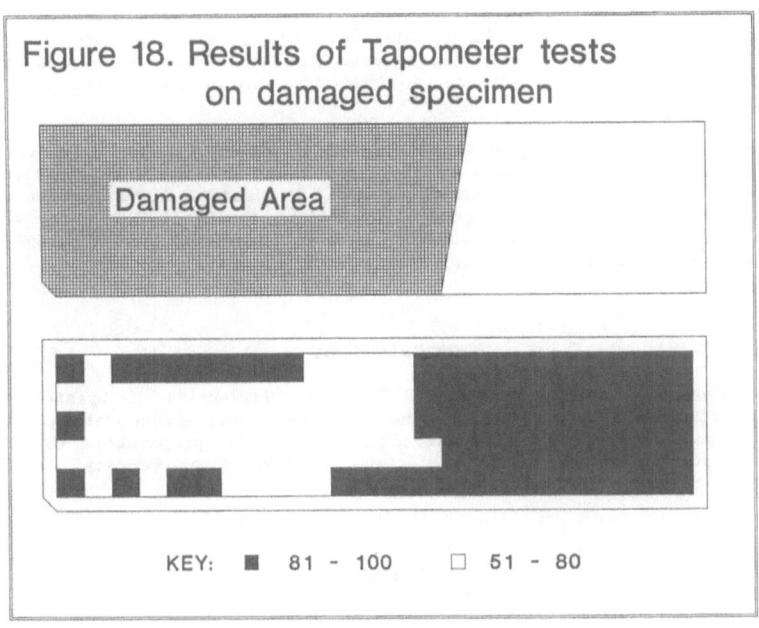

Figure 18. Results of Tapometer tests on damaged specimen

Damaged Area

KEY: ■ 81 - 100 □ 51 - 80

CONCLUSIONS

As a result of this investigation, the following conclusions can be drawn:

o Disbonded regions of joints can be detected quite accurately with the Tapometer.

o Large voids (typically greater than 10 mm diameter) can be readily detected.

o Bondline inclusions (even up to 2 mm thick and 10 mm diameter cannot be detected with any confidence.

o Varying bondline thicknesses in the range 0 - 3 mm can be tested without re-calibrating.

o The technique has been shown to work on substrates with simple curvature.

o Varying cross-sections in joints can be detected reasonably well.

More generally, it is has been seen that assessment of bond quality cannot be made from an individual test result. Instead, a series of tests must be carried out over the whole joint area to identify defective regions. This can obviously make interpretation of the results difficult. Also, in many cases, anomalous readings are obtained at test sites close to the edge of the specimens. This is of particular concern to the automotive industry because typical flange widths (say 15 mm) are effectively all "edge".

It should also be stressed that the current shape and size of the Tapometer's tapping head make it difficult to use automotive structures and that further work is required to establish the effect of tapping onto a finished (eg painted) surface on both the results and on the surface finish.

In summary, since the Tapometer is only able to detect gross defects and there are a number of unresolved problems in its use and interpretation, it looks unlikely that this technique could be used as a quality control method in the manufacture of bonded automotive structures. However, with further development, it could possibly be used for assessing the extent of bondline damage following impact loading.

ACKNOWLEDGEMENTS

The authors would like to thank R D Adams (University of Bristol), P Cawley (Imperial College, London) and P Walkden (Rolls-Royce MatEval Ltd) for help during the experimental work and in the preparation of this manuscript. Thanks must also be given to Ford Motor Company Ltd for allowing these results to be published.

REFERENCES

1) Wang, T.T., Ryan, F.W. and Schonhorn, H., 'Effect of bonding defects on shear strength in tension of lap joints having brittle adhesives', Journal of Applied Polymer Science, Volume 16, 1972, pp 1901-1909.

2) Segal, E. and Rose, J.L., 'Non-destructive testing techniques for adhesive bond joints' in 'Research techniques in Nondestructive Testing IV, Academic Press, London, 1980, pp 275-316.

3) Schliekelmann, R.J., 'Non-destructive testing of adhesive bonded joints', AGARD Lecture Series 102, 1979, pp 8-1 - 8-37.

4) Adams, R.D. and Cawley, P., 'A review of defect types and nondestructive testing techniques for composites and bonded joints', NDT International, Volume 21, Number 4, August 1988, pp 208-222.

5) Adams, R.D. and Cawley, P., 'Vibration techniques in non-destructive testing', in 'Research techniques in non-destruction testing VIII', edited by R.S. Sharpe, Academic Press, London, 1985, pp 303-360.

6) Cawley, P. and Adams, R.D., 'The mechanics of the coin-tap method of non-destructive testing', Journal of Sound and Vibration, Volume 122 Number 2, 1988, pp 299-316.

7) The MatEval Tapometer operator's manual, Published by Rolls Royce MatEval Ltd, Warrington, UK.

3

THE BLISTER TEST AS A MEANS OF EVALUATING THE PROPERTIES OF RELEASE AGENTS

B.J. Briscoe and S.S. Panesar*
Department of Chemical Engineering, Imperial College, London SW7 2BY

*current address: BP Research Centre, Sunbury-on-Thames

ABSTRACT

The paper describes the use of the "blister test" to evaluate the influence of surface topography upon the adhesion of a poly(urethane) to a range of steel substrates. The steel substrates have been selected with a range of surface roughness and quantified using a stylus profilometer. In addition, a number of these surfaces have been coated with a selection of materials which act, more or less, as release agents.

The early part of the paper concentrates upon the experimental method and considers, amongst other things, the effect of crack migration and the selection of an appropriate analytical expression for the computation of the peeling energy. The later part of the paper describes the separate and combined effects of the roughness and these release agents. The main emphasis of the paper is to demonstrate the general value of the blister test as a means of quantifying the effect of release coatings on adhesion.

1. INTRODUCTION

This paper considers two general problems in the context of the reduction of the adhesion of a polyurethane to a solid substrate. One part of the paper deals with the application of the so called blister test to enable the acquisition of reliable experimental data to assess the magnitude of the peeling energy. This part of the paper deals with such things as the experimental method and the analysis of the experimental data. It is shown that, within the confines of the present work, that a good and reliable estimate of the peeling energy may be obtained from an instrumented blister test. This part of the paper considers the resolution of certain experimental difficulties and the choice of an appropriate analysis of the measured quantities to provide an estimate of the peeling energy.

The second part of the paper provides a range of experimental data and a first order interpretation of their significance. The study of two interfacial variables factors are considered as examples of the application of the technique. There are the action of release agents and the effects of counterface topography and also a combination of the two. As an example, two effective release systems are chosen based upon carboxylic ester wax suspensions and a model multiphase silicone based release system. These systems are very effective in reducing the measured adhesion. In addition, the influence of surface topography is considered first in the case of the uncoated counterfaces. We describe an unexpected, but rationalised result, that increasing roughness, based upon a variety of topographical criterion, reduces the adhesion. This effect is envisaged as being due to ineffective wetting of the polyurethane on the rougher substrates. The combined effect of counterface roughness and release agent causes an apparent marked increase in adhesion. We have interpreted this trend as being due to a reduction in the effective roughness of the counterface and hence a better wetting of the substrate.

2. EXPERIMENTAL

2.1 Preparation of Polyurethane

The polyurethane used in this study was based on a polycaprolactone ethylene glycol adipate polyester and a diphenyl methyl -4'-4'- diisocyanate, which were supplied by B & T Polymers (Chester, England). The same polymer has been used in the other studies by the present authors, details of which have been published previously [1,2,3]. The polymer was prepared by mixing the polyol and the diisocyanate in the ratio of 8.5:1 (by weight). The mixture was then degassed in low vacuum (10^{-1} torr) after which it was cast onto a substrate (section 2.2) of known surface topography. A simple PTFE ring was used to contain the viscous mixture whilst curing took place; typically curing was assumed to be completed after two hours.

The cured polyurethane was peeled from the substrate by the blister test procedure which is described later. The polyurethane was presumed to have conformed to the surface topography of the substrate, assuming that the liquid polyurethane had completely wetted the substrate surface.

2.2 Counterfaces and their Preparation

The basic substrate was fabricated out of mirror finished stainless steel, 2mm thick and 200mm square. The centre of the plates contained the necessary hole for the connection to the pneumatic gas supply (see later). The mirror finished stainless substrate was abraded with different grade emery paper to vary the surface roughness. The plates were cleaned with 1,1,1 trichloroethane and 'Analar' acetone prior to use. The topographical characteristics are described in section 5.1.

2.3 Release Agents

Two types of release layers were used:

(a)   stearyl stearate suspensions;

(b)   silicone multiphase systems, which consisted of Wacker MK (supplied by Wacker Silicones), a cross-linked methyl resin which was combined with a PDMS silicone fluid of 500 cps approximate viscosity in the ratio of 1 to 5.

Both coatings were deposited by spraying from dilute solutions (concentrations 1 gm^{-2}) using an air atomiser positioned vertically over the steel substrate. The steel substrate was moved at a constant velocity beneath the spray and a series of screens were used to ensure uniform deposition. 1,1,1 trichloroethane was the spraying solvent, and the coating produced appeared to be 'dry' (free of solvent) a few seconds after the deposition. The coated substrates were stored in a vacuum dessicator which was continuously pumped for some hours before the samples of polyurethane were cast against the coatings. Typically, the coating thicknesses used were 1 μm.

2.4 Experimental Method and Apparatus

The apparatus and experimental techniques used to measure the adhesion fracture energy of the polyurethane-steel interfaces have been described previously [1]. It involved the use of the so-called blister test. A brief description of the apparatus is given here. A schematic diagram of the complete apparatus is shown in Figure 1.

The debonded hole of diameter 2a, shown in Figure 2, was sealed with a close-fitting metal plug which was arranged to be flush with the metal surface. The polyurethane was cast against the metal substrate. After the polymer had been cured against the steel substrate the blanking plug was removed (Figure 1c) and the hole in the steel plate attached to a N$_2$ gas line which included a flowmeter, a gas supply and a pressure transducer. A counterbalanced transducer and a high speed camera were positioned vertically over the unbonded polyurethane region. In some experiments the camera (Vinten MK3) was positioned to monitor the profile of the blister. The camera was operated at 48 frames per sec. The processed microfilm was analysed with an Emulsion Image Analyser. The debonded area was pressurised at a constant flow rate controlled by the flow controller. A pressure transducer was used to monitor the pressure, P, as a function of time, and digitial pressure readings were stored in a micro-computer. Simultaneously, the relative height, H, of the debonded region was sensed by the displacement transducer and the A/D converter transmitted the values of H as a function of time into the same microcomputer. By these means, P and H were measured as a function of the lapsed time, t; Figure 3 shows a typical set of data.

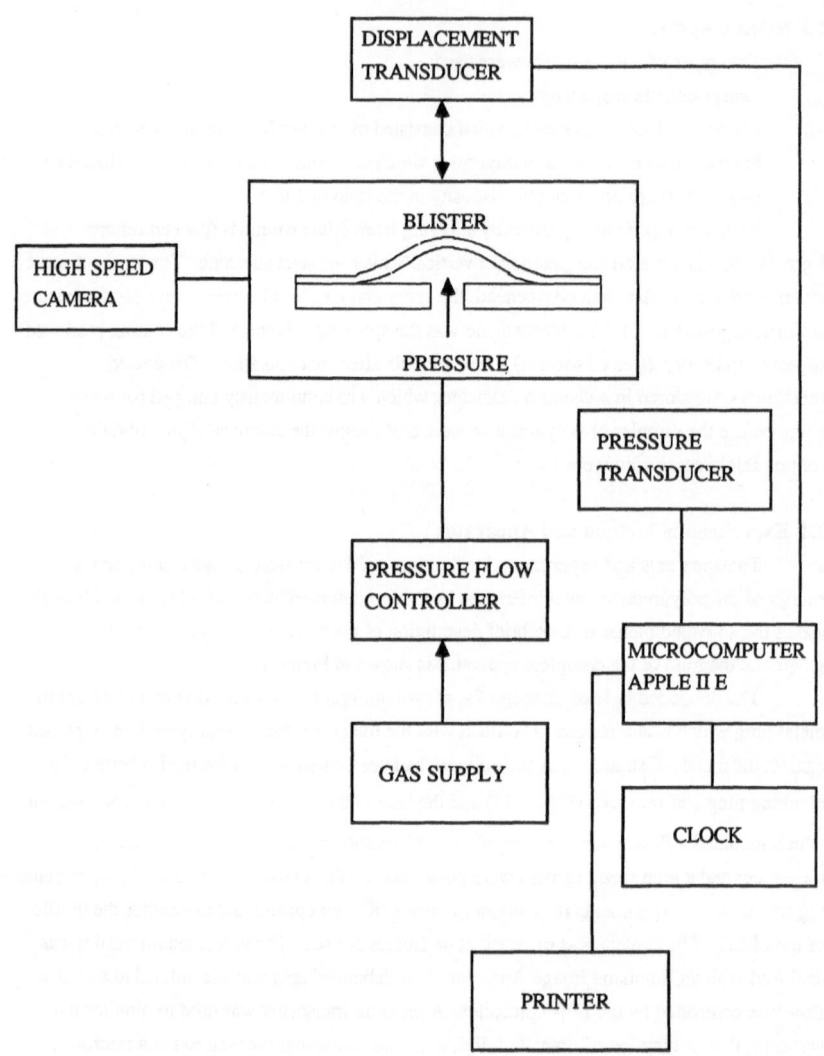

FIGURE 1 A schematic diagram of the experimental arrangement used to obtain
P, H and a as a function of time

41

FIGURE 2   A diagram of a blister test

(a) prior to pressurisation of polyurethane/steel interface; a is the entry hole radius

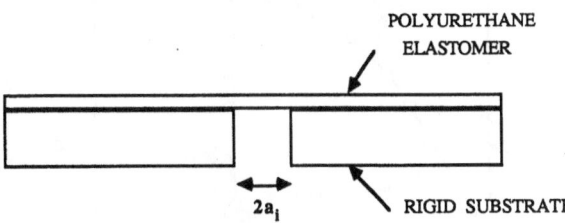

(b) at blister formation of polyurethane/steel interface

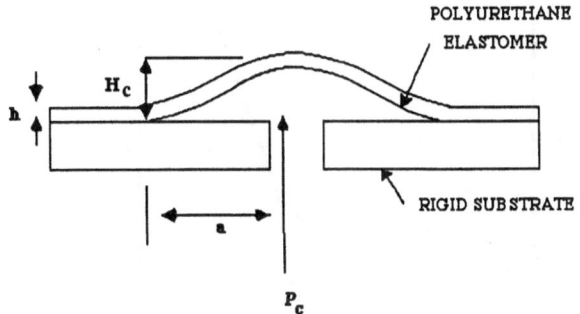

(c) description of plugging arrangement for the debonded entry hole in the substrate prior to casting of polyurethane elastomer

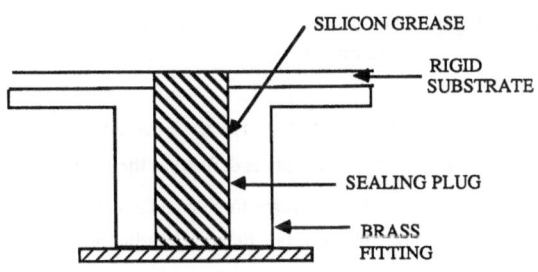

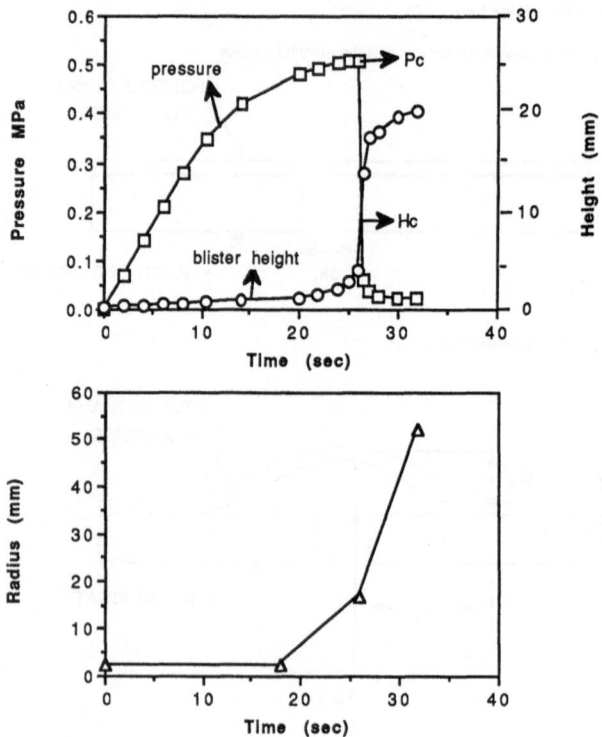

FIGURE 3   The transducer reponse for blister height, H, pressure, P, and radius, a as a function of time.

At a critical condition, the pressure reaches a maximum and the crack begins to accelerate; at this point, the pneumatic pressure is defined as the critical pressure, P_c (Figure 3). Eventually the peeling condition becomes catastrophic and a rapid increase in velocity is observed. Figure 3 also shows the corresponding crack radius as a function of time. Likewise, H is defined as a critical height H_c. The fact that there is some peeling of the interface regions adjacent to the debonded region prior to the development of the maximum pressure, P_c introduces a serious complication in the subsequent analysis as the value of the debonded radius at P_c (or H_c) is required. The extent of this premature peeling depends upon the system but Figure 3 is a typical example. This point is discussed later but we assume for the moment that the required debonded area is a crack of radius a_c; $a_c > a_i$, where a_i is the initial debonded radius or substrate port hole radius.

Two experimental approaches were used. The normal method adopted in most blister tests is to attempt to fix the critical debond radius, a_c, as the initial geometry of the debonded region which is attached to the pressure source. This we describe as the predefined radius method. The value of a_c, the crack radius at the rupture, is presumed to be close to the value of the radius of the hole in the plate, a_i. The second method, the preblown radii technique, proposed by Williams [4], Anderson et al [5] & Takashi et al [6], is less common and involves progressively increasing a_c by preblowing the blister to values larger than the initial value of a_i. This experiment involves carefully rupturing the interface at the original entry point to produce a series of larger blisters. A blister is formed and then the pressure is released and the blister is collapsed. The experiment is repeated several times, allowing slight growth in the blister size each time. The experimental problem of defining the appropriate value of the radis, a, was simply achieved by examining the surface of the polyurethane immediately after rupture. A series of readily detectable concentric rings was apparent on the polyurethane and their radii were easily measured. The rings fade rapidly with time and correspond to highly deformed rubber produced at the termination of crack motion; typically, these distortions are about 300 μm in height. High speed photography was also used to estimate a_c.

The crack velocity at the contact condition is given to a good approximation by equation (1), a relationship which has been derived in earlier work [1].

$$v = P_c \, F \, t_c \, (d(1/P_c)/\pi \, a_c \, H_c \qquad (1)$$

where the subscript c indicates that the values of the parameters are those at the critical rupture condition. F is the constant gas flow rate, t is the time lapsed, and H is the blister height. The flowmeter supplied a constant nitrogen gas flow which ranged from $5.1-0.01 \times 10^{-4} \text{m}^3\text{s}^{-1}$ and the pressure transducer was accurate to ±550 Pa and blister heights were recorded better than ±50 μm. Data were typically recorded every 0.125s.

As reported earlier [2], the blister/substrate interface undergoes stable crack growth prior to catastrophic failure. Ignoring the existence of stable crack growth prior to catastrophic failure leads to incorrect a_c values and substantial errors during the calculation of values of Γ and v. The use of high speed photography, which is expensive, seems to be the only precise means of measuring the value of a_c. However, a simple correction procedure is available in the form of equation (2), which presumes that H is proportional to a, at rupture.

$$\frac{a_c}{H_c} = \frac{\Delta a}{\Delta H} = K \qquad (2)$$

where ΔH is the difference in blister height and Δa is the difference in radii between the initiation and the termination of each preblown blister propagation event. Δa is readily measured as the distance between the ring marks, whereas ΔH is obtained for the H vs t profiles (Figure 3); hence K. H_c corresponds to P_c and hence the corresponding a_c can be estimated from the preblown radii blister technique after K is computed. Further details of this approach are published elsewhere [2].

Separate simple peel tests were performed using the 90° peel test configuration; this consisted of a 50mm wide and 5mm thick polyurethane strip, which was cast onto the stainless steel plate. One end of the strip was incorporated into a fitting which had a hook and weights attached to it. The relative peeling times and the length of strip peeled were measured using a stop watch and a travelling microscope. The fracture energy, Γ, was computed by the methods described by Kendall [7] and the crack velocities were also calculated.

3. BLISTER TEST ANALYSIS

The blister test was originally suggested as a means of measuring the adhesion of elastic solids by Dannenberg [8], and developed by Malyshev & Salganik [9]. The present theoretical understanding of this technique is largely due to the work of Williams and co-workers [4,5,10-18]. Recently, Andrews & Stevenson[19], Wronski & Parry [20], Updike [21], Takashi *et al* [6], Yamazaki & Takashi [22] and Hinkley [23] have extensively applied and evaluated this technique for studying the adhesion of solids. Gent & Lewandowski [24] have more recently used it for studying the adhesion of membranes to rigid plates. Napolitano and Senturia [25] have suggested the use of a "constrained" blister test which allows the testing of strongly adherent thin films. Table 1 lists the equations currently available from the literature which allow the calculation of a value of the peeling energy, $\Gamma(v)$, from the parameters P_c, a_c and certain subsidiary parameters such as Young's modulus, E, the Poisson's ratio, υ and the specimen thickness, h. Table 1 shows that a distinction is often made between two regimes determined by the ratio between elastomer thickness, h, and crack radius, a. When h/a is larger than (h/a > 5) the appropriate solution is termed as infinite medium solution. For 5 > h/a > 0 the system is described as a thick plate or membrane and here a number of options are available. They are generally of two types depending upon whether the approximations of exclusive bending or stretching deformations are adopted to compute the stored elastic strain. The choice of the appropriate solution often presents serious difficulties and is a major source of potential error. They all assume elastic behaviour with the exception of Yamazaki & Takashi [22] who have included a viscoelastic response. There are two broad classes which may be distinguished according to the mode used to describe the storage of elastic deformation energy. For example, Andrews & Stevenson [19] considered the elastic bending energy contribution whilst Hinkley [23], Takashi et al [6] Gent & Lewandowski [24] and Briscoe & Panesar [2] computed the stretching energy. The choice of the appropriate deformation model and hence the necessary analytical equation, is facilitated, in principle, by investigating the shapes of the blisters. The blister shapes produced by two modes of deformation are characteristically different and predictable. The expected inter-relationships between H_c and a_c are different. For stretching only the spherical cap geometry approximates to:

$$(2R - H_c) \, H_c = a_c^2 \qquad (3)$$

where R is the cap radius

Alternatively the membrane approximation [26]

$$H_c = 0.597 \, a_c \, [P_c \, a_c^4 / Eh]^{1/3} \qquad (4)$$

For bending only, assuming $h < a/2$ and that $H_c < h/2$:

$$H_c = \frac{P_c a_c^4}{64D}$$

where D is $Eh^3/12(1-\upsilon^2)$

For bending at large strains, stretching in the middle has been incorporated [27] and:

$$H_c + 0.58 \, \frac{H_c^3}{h^2} = \frac{P_c a_c^4}{64D}$$

Figure 4 shows H_c against a_c for the two extreme models based on bending and stretching and also selected experimental data for the present study.

TABLE 1
Various Relations Available for Computing Fracture Energy From Blister Test Data

Author Equation	Type of Storage Energy Model Used
Dannenberg	Total work

$$\Gamma = P_c \, \Delta V$$

Williams (a) Infinite Media	Total work

$$\Gamma = \frac{P_c^2 a}{E} \frac{2(1-v^2)}{\pi}$$

(b) Thin Specimen   Bending energy

$$\Gamma = \frac{P_c^2 a}{E} \left(\frac{a}{h}\right)^3 \frac{3(1-v^2)}{32}$$

Andrews *et al*   Bending energy

$$\Gamma = \frac{P_c^2 a}{E} (1-v^2) \left[\frac{3}{32}\left[\left(\frac{a}{h}\right)^3 + \left(\frac{a}{h}\right)\frac{4}{1-v}\right] + \frac{1}{\pi}\right]$$

Hinkley   Stretching energy

$$\Gamma = \frac{P_c H_c}{4}$$

Broutman & McGarry   Bending energy

$$\Gamma = \frac{P_c H_c}{8} \times n^*$$

Takushi *et al*

$$\Gamma = \left[\frac{P_c^4 \, a^4}{18.2Eh}\right]^{\frac{1}{3}}$$

Gent & Lewandowski (a) Radius   Stretching energy

$$\Gamma = \left[\frac{P_c^4 \, H_a^4}{17.4Eh}\right]^{\frac{1}{3}}$$

(b) Height $\Gamma = 0.65 \, P_c \, H_c$ Stretching energy

Present paper   Stretching energy

$$\Gamma = \left[\frac{P_c^4 H_c^4}{576Eh \,)1-v)^2}\right]^{\frac{1}{3}}$$

* is a constant, its value for blister experiments $\cong 2$

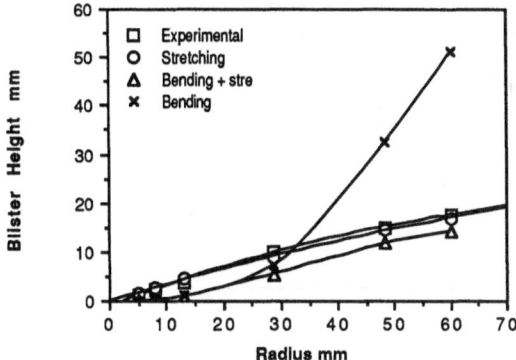

Figure 4: Plot of Blister height against debond radius for experimental data, stretching model, stretching plus bending model and bending model.

None of the predicted shapes is completely adequate in describing the present experimental data over the whole range. However, for the experimental range of the crack radii, the elastic stretching case is the closest, and an energy storage model based upon stretching is most appropriate. Therefore, an analysis based on stretching, which is fully described elsewhere [2], is used to calculate fracture energy of the polyurethane-steel interface studied in the present work. We have used the relationship [2].

$$\Gamma = \left[\frac{P_c^4 a_c^4}{576(1-\upsilon)^2 Eh} \right]^{\frac{1}{3}} \tag{5}$$

4. RESULTS

4.1 Uncoated Smooth Polyurethane-Steel Interfaces

Figure 5 shows typical data for Γ as a function of the estimated peeling velocity. The computed values of Γ are a strong function of the corresponding crack velocity which is in part determined by the imposed flow rate.

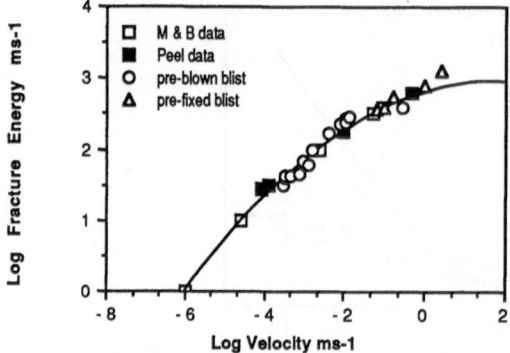

Figure 5: Plot of log fracture energy against log velocity for Maugis & Barquin data, peel data, pre-blown and pre-fixed blister data.

Of more significance however is the good agreement observed between the derived values of

Γ based on the stretching analysis (equation 5) for the blister data and the sample peel data obtained for the rupture of smooth polyurethane-steel interfaces (Figure 5). Both sets of data are also in close agreement with the results represented by Maugis & Barquin [28]. These authors studied a polyurethane/glass system using a 90° peel test. Figure 5 shows a comparison of these three sets of data, which shows that the blister analysis based on

stretching (equation 5) provides not only an accurate quantitative prediction of the value of Γ

but also its variation with velocity for the polyurethane adhesive layer. The value of $\Gamma(v)$ appears to vary as $v^{0.6}$ [28].

4.2 Smooth Coated Polyurethane/Steel Interfaces

Two types of release coating systems have been studied in conjunction with the blister test; the ester and silicone based systems. The effectiveness of these release coatings has been described in detail previously [1,3]. Selective release layers for the two systems are described; (a) a stearyl stearate suspension and (b) a multiphase silicone based system (Wacker MK resin + PDMS (500 cps)) 1 : 5. Figure 6 shows a plot of fracture energy against

the crack velocity for uncoated and release layer coated interfaces. Table 2 shows the Γ values at 1 ms^{-1} velocities for the coated and uncoated interfaces. It is evident that the stearyl stearate suspension coating decreases the fracture energy by 40 fold, whilst the silicone multiphase coating decreases the fracture energy by 600 fold. This attenuation of the adhesion energy is apparently due to the formation of friable, marginally bonded particle based structures which are sufficiently durable to withstand the polymerisation process but are readily ruptured or fractured during the subsequent peeling of the interface.

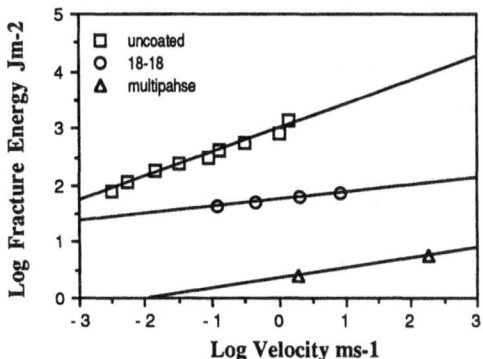

Figure 6: Plot of log fracture energy against log velocity for uncoated interface, stearyl stearate suspension and silicone multiphase coated interfaces.

TABLE 2

Fracture energy, G and normalised fracture energy Γ/Γ_0 for coated and uncoated interfaces

COATINGS	Γ (JM^{-2})	Γ/Γ_0
VIRGIN	1000	1
18-18 SUSPENSION	54.84	0.0548
SILICONE MULTIPHASE	2.227	0.0022

5. SURFACE ROUGHNESS

5.1 Surface Characterisation

The surface roughness characteristics of the stainless steel substrates of varying roughness were measured using a Rank Taylor Hobson Talysurf stylus instrument interfaced with an on-line microcomputer. The surface profiles were recorded on 2024 data points taken at a 2 μm sampling interval using a high resolution stylus (radius 2 μm x 5 μm). Several transverse profiles were obtained for stainless steel substrates; (i) mirror finished, (ii) lathed and (iii), (iv), (v) mirror finished surface roughened by different grades of emery paper. The surface profiles detected by the stylus were digitised and processed by the microcomputer. Hence, a number of topographical parameters were derived which are listed in Table 3.

TABLE 3

Topographical parameters

(a)

	CLA μm	RMS μm	SD(σ) μm	Skewness	Kurtosis	Peak curvature	β*	σ/β*	(σ/β*)$^{1/2}$
1	0.044	0.059	0.06	-1.00	5.43	0.01	108.7	0.001	.023
2	0.224	0.292	0.292	-0.56	4.36	0.062	4.34	0.067	0.25
3	0.170	0.218	0.218	-0.45	3.54	0.066	13.97	0.015	0.12
4	0.272	0.353	0.353	-0.63	4.06	0.057	10.65	0.033	0.18
5	0.410	0.549	0.549	0.386	5.15	0.047	55.43	0.009	0.099

(b) measured 90° to (a)

	CLA μm	RMS μm	SD (σ) μm	Skewness	Kurtosis	Peak curvature	β*	σ/β*	((σ/β*)$^{1/2}$
1	0.0452	0.058	0.059	-0.17	3.53	0.011	130	0.001	0.021
2	0.116	0.142	0.143	-0.231	2.66	0.006	153	0.001	0.029
3	0.173	0.223	0.223	-0.25	3.67	0.063	7.6	0.029	0.17
4	0.259	0.327	0.328	-0.38	3.12	0.065	6.52	0.050	0.22
5	0.463	0.601	0.601	0.009	4.71	0.043	48.4	0.012	0.111

5.2 Topographical Analysis

In general two functions are necessary to describe the topography of a surface [29]. One may be thought of as describing the vertical features, this is usually taken as the distribution of ordinate heights with respect to some datum level. For a Gaussian distribution of ordinate heights this may be defined by their standard deviation σ. σ is approximately equal to the centre line average for Gaussian distribution. The other function which is required describes the horizontal distribution of the roughness. A useful way of providing this conformation is by means of the auto-correlation function or peak curvature distribution. An exponential correlation function gives a good approximation to many practical randomly prepared profiles [30], and this can be specified by the autocorrelation distance $\beta*$, the spacing at which the autocorrelation function decreases to 1/e. Alternatively, a mean value can be obtainable from peak curvature distribution directly, β, which is the radius of curvature of the asperity tip. According to Greenwood & Williamson [31], β is directly related to the standard deviation of ordinate heights and the correlation distance:

$$\beta \, \alpha \, \frac{\beta_*^2}{\sigma}$$

$$\sigma \, \alpha \, \beta \, \sigma*$$

where $\sigma*$ is the root mean square value.

5.3 Effect of Roughness on Uncoated Interfaces

Figure 7 shows a plot of log Γ against log v for unadulterated polyurethane/steel interfaces with the varying surface roughness described in Table 3.

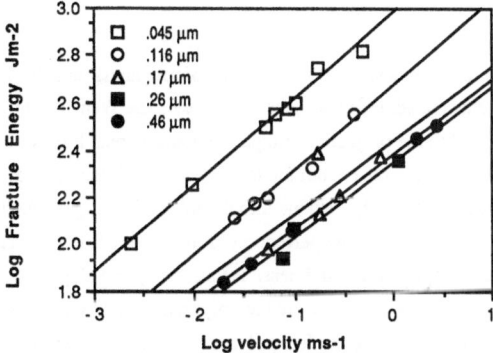

Figure 7. Plot of log fracture energy against log velocity for uncoated interfaces of varying roughness

Figure 8 shows a plot of Γ at 1 ms^{-1} velocity obtained from Figure 7 as a function of the centre line average (C.L.A.) values of the steel substrate. Clearly, the peeling energy decreases as the surface roughness (C.L.A. values) are increased from 0.045 to 0.026. A further increase in surface roughness, i.e. >0.26 C.L.A. values, results in invariant peeling energies. However, the C.L.A. values are only representative of the vertical distribution of the asperities and do not take account of the horizontal features of the roughness. Inspection of Table 3 however indicates that the fracture energy at 1ms^{-1} is also a decreasing function of

β^* which is generally regarded as a good description of the horizontal texture of the topography.

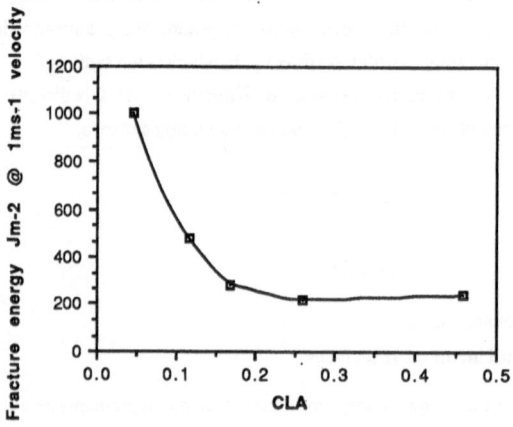

Figure 8: Plot of fracture energy obtained from Figure 7 at 1ms^{-1} constant velocity vs CLA values.

The data are at odds with the accepted wisdom of the influence of roughness on adhesion; normally one anticipates an increase in adhesion perhaps scaling with the Wenzal Ratio. The Wenzel Ratio is the ratio of the real to the projected or apparent contact area. It is possible that the roughness features produce stress raisers but the trends are more likely due to imperfect wetting or inhibited entry of the polyurethane into the microcavities generated by the roughness. It is reasonable to assume that the perfection of the wetting is controlled by, amongst other things, the average slope of the roughness. Relatively gentle slopes will not greatly hinder the dynamic wetting of the roughness by the polyurethane and vice versa. Similarly, the necessary Kelvin pressures for complete polyurethane intrusion will be less for the lower slopes. The data in Table 3 provides a number of estimates of the mean slopes as

sensed by the stylus and the associated measuring system. Figure 9 compares the adhesion
with the ratio of $\sigma/\beta*$ which has consistently been regarded [32] as a good measure of the
mean slope of the asperities. Equally β and $(\sigma/\beta*)^{1/2}$ are arguably reasonable descriptions
of the slope. It is apparent from Figure 10 that the adhesion decreases with the inverse of the
slope of the asperities. This is consistent with a model of imperfect wetting of rough
surfaces.

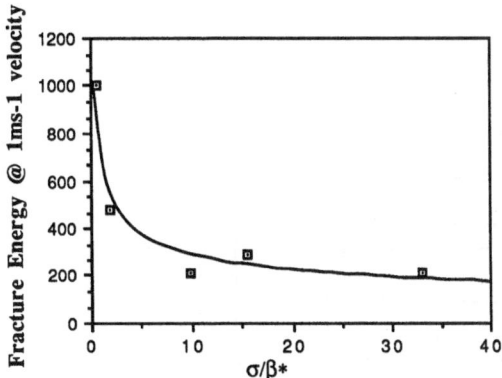

Figure 9: Plot of fracture energy obtained from Figure 7 at $1ms^{-1}$ constant velocity vs $\sigma/\beta*$

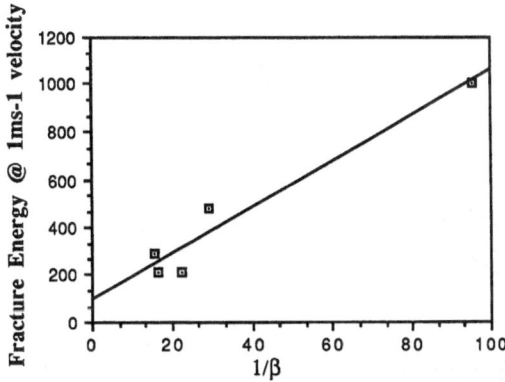

Figure 10: Plot of fracture energy obtained from Figure 7 at $1ms^{-1}$ constant velocity vs $1/\beta$

6. EFFECT OF ROUGHNESS ON THE COATED POLYURETHANE/STEEL INTERFACES

Two types of coatings (a) stearyl stearate suspension and (b) multiphase (Wacker MK resin + PDMS (500 cps) (1:5) were investigated with polyurethane/steel interfaces of varying roughness. Figure 11 and Figure 12 show plots of log Γ against log v for the two release coatings.

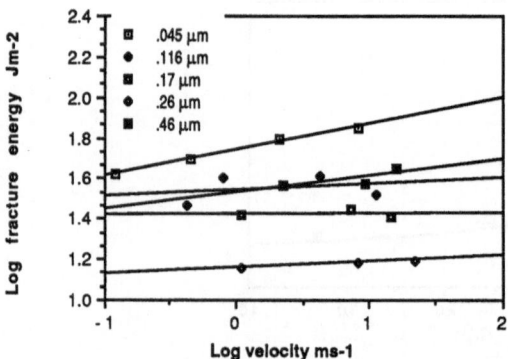

Figure 11: Plot of log fracture energy against log velocity for stearyl stearate suspension coated interfaces of varying roughness

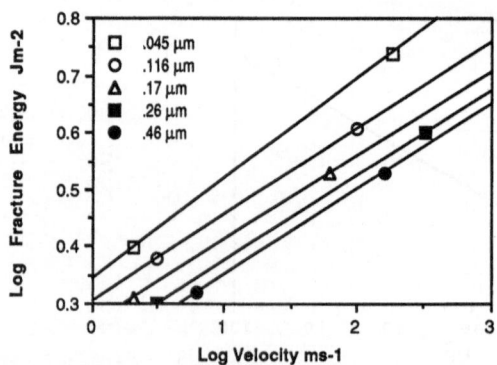

Figure 12: Plot of log fracture energy against log velocity for multiphase silicone coated interfaces of varying roughness.

Figures 13 and 14 shows a plot of the normalised fracture energies, $\Gamma/\Gamma_o{}^*$, obtained at 1 ms^{-1} from Figure 11 and Figure 12 as a function of the C.L.A. values for the two release coatings. $\Gamma_o^{r\ *}$ is the fracture energy of the corresponding uncoated rough counterface.

Unlike the uncoated system the normalised fracture energies for both of the coatings increase with the increased surface roughness, i.e. in this case the C.L.A. values.

These data are for a mean coating thickness of ca. 1μm. Similar data are obtained for other thicknesses but thinner coatings produce greater adhesion [3]. The interesting result is that the release coatings increase the adhesion and in a simplistic way we may regard this action as a "smoothing" the rougher substrates. Doubtless their true action is more complex in nature.

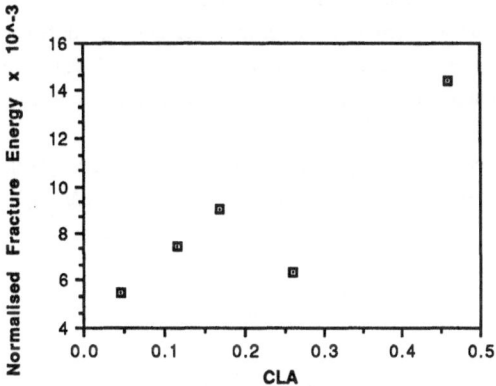

Figure 13: Plot of normalised fracture energy against CLA values for stearyl stearate coated interfaces

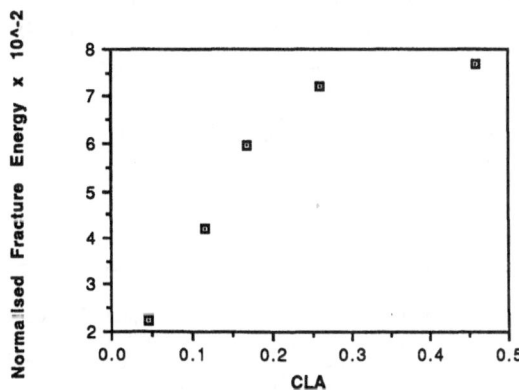

Figure 14: Plot of normalised fracture energy against CLA values for multiphase silicone coated interfaces.

7. CONCLUSION

This paper has reviewed certain facets of the use of the blister test as a means of quantifying the peeling energy of poly(urethane) steel substrates. We conclude that with the proper choice of experimental method and the subsequent analysis a good and reliable estimate of the peeling energy may be obtained. We have also shown how this energy is influenced by the crack velocity, the counterface roughness, the presence of release agents and the combination of the latter two.

At a physical level of understanding we have offered brief evidence of the effective value of multiphase release agent systems. Certain multiphase systems provide very effective reductions in adhesion. The introduction of roughness in the present system greatly reduces adhesion. The roughness has been characterised by established procedures. More roughness - less adhesion. We have interpreted this result as arising from indifferent wetting of the poly(urethane) into the interstices of the steel substrate roughness. With release agents which effectively smooth the substrates we interpret the increase in adhesion as being due to better wetting.

Whilst the various interpretations of the affects of the coating and the roughness, solely and combined, are open to question, the paper illustrates the values of a careful application of this blister test to estimate the adhesion and abhesion of poly(urethane) contacts.

8. REFERENCES

1. Briscoe, B.J. and Panesar, S.S., 1986, J. Phys. D. Apply. Phys., **19**, 841.
2. Briscoe, B.J. and Panesar, S.S., 1989, Submitted to Proc. Roy. Soc. (London).
3. Briscoe, B.J. and Panesar, S.S., 1988, J. Adhesion Sci.Tech., **2**, 4, 287.
4. Williams, M.L., 1972, J.Appl.Polym.Sci., **3**, 1.
5. Andreson, A.P., Bennet, S.J. and Devries, K.L., 1972, "Analysis and Testing by Adhesive Bonds", Academic Press, NY.
6. Takashi, M. and Yamazaki, K., 1987, Proc. Jpn. Congr. Mat. Sci., 960.
7. Kendall, K., J. Phys. D: Appl. Phys., 1973, **6**, 1782.
8. Dannenberg, H., 1961, J. Appl. Sci. Polym. Sci., **5**, 125.
9. Malyshev, D.M. and Salganik, R.L., 1965, Int. J. Fract. Mechs., **1**, 11.
10. Williams, M.L., 1969, J. Appl. Polym. Sci., **13**, 29.
11. Williams, M.L., 1970, J. Appl. Polym. Sci., **14**, 1121.
12. Williams, M.L., 1972, J. Appl. Polym. Sci., **4**, 307.
13. Williams, M.L., 1973, J. Appl. Polym. Sci., **5**, 81.
14. Williams, M.L., 1978, J. Appl. Polym. Sci., **9**, 145.
15. Anderson, A.P., Devries, K.L. and Williams, M.L., 1974, J. Col. Interf. Sci., **47**, 600.
16. Anderson, A.P., Devries, K.L. and Williams, M.L., 1974, Int. J. Fract., **10**, 565.
17. Bennett, S.J., Devries, K.L. and Williams, M.L., 1974, Int. J. Fract., **10**, 33.
18. Burton, J.D., Jones, W.B. and Williams, M.L., 1970, Proc. Soc. Rheol., California.
19. Andrews, E.H. and Stevenson, A., 1978, J. Mat. Sci., **13**,1680.

57

20. Wronski, A.S. and Parry, T.V., 1980, "Adhesion 5", edited by Allen, K.W. (Applied Science Publishers, Lond.), 1.
21. Updike, D.P., 1976, Int. J. Fract., **12**, 815.
22. Yamazaki, K. and Takashi, M., 1978, Proc. Jpn. Cngr. Mat. Sci., 955.
23. Hinkley, J.A., 1983, J. Adhesion, **16**, 115.
24. Gent, A.N. and Lewandowski, H., 1987, J. Appl. Poly. Sci., **33**, 1567.
25. Napolitano, M.J. and Senturia, A., 1987, Paper presented at Amer. Adhs. Soc., Williamsberg.
26. Hencky, H., 1915, Z. Math. Phys., **63**, 311.
27. Timoshenko, S., 1981, 'Strength of Materials", Part II, Van Nostrand Reinhold, NY, 101.
28. Maugis, D. and Barquins, M., 1978, J. Phys. D: Appl .Phys., **11**, 1989.
29. Archard, J.F., 1974, Trib. Int., 213.
30. Whitehouse, D.J. and Archard, J.F., 1970, Proc. Roy. Soc. (London), **A316**, 97.
31. Greenwood, J.A. and Williamson, J.P., 1966, Proc. Roy. Soc. (London), **A295**, 300.
32. Halliday, J.S., Proc. Inst. Mech. Engr., 1955, **169**, 777.

4

THE USE OF PRIMERS IN BONDING POLYPROPYLENE

A. BEEVERS & T. NORRIS

Oxford Polytechnic
Joining Technology Research Centre

INTRODUCTION

Polypropylene finds use in both domestic and industrial applications for packaging, automotive trim and for components of washing machines etc. Its mechanical properties, low density and corrosion resistance make polypropylene an attractive substitute for many established materials, but it is difficult to bond this polymer using structural adhesives. This prevents polypropylene from being more widely used.

Surface treatments such as acid etching, flaming and corona discharge have been used for many years to make polypropylene receptive to paints and glue. Organic primers such as chlorinated polypropylene have also been used for decorative coatings but not for load-bearing joints. Recent research has shown that the use of these primers enables polypropylene to be strongly bonded to other materials using structural adhesives.

ADHESIVE BONDING OF POLYPROPYLENE

Polypropylene surfaces are chemically inert : they consist entirely of hydrocarbon macromolecules and have no functional groups with which a polar adhesive, ink or solvent can interact. However, polypropylene, unlike polyethylene, contains 'tertiary' C-H bonds which are more susceptible to oxidation than primary or secondary C-H bonds. This oxidation may be caused by chemical reagents such as chromic acid or potassium permanganate; or it may be caused by molecular fragments in an electric discharge or a flame.

The surface treatments used on polymers have been reviewed by Briggs [1] and Wu [2], while detailed reports have been published on the use of flames [3,4], chromic acid [5,6], permanganate [7] and electric discharges [8,9,10]. Essentially all these treatments attack the surface of polypropylene and replace C-H bonds by C-O, either as carbonyl or hydroxyl functions. Discharge treatment may produce crosslinking, while etching solutions may leave chemically bound metal species on the surface.

CHLORINATED POLYPROPYLENE PRIMERS

Chlorinated polypropylene is a slightly rubbery thermoplastic material which can be prepared by solution reaction or suspension reaction between chlorine gas and polypropylene. Suspension reactions give more highly crystalline products which show greater adhesion (but are higher-melting and less soluble) than solution-chlorinated polymers of equal Cl content [11].

Propylene polymer (or copolymers) can be chlorinated in solution at about 150°C, or as a suspension in water at 90°C. An organic peroxide compound or UV light irradiation is used to promote reaction between chlorine and polymer. Another method involves kneading molten polymer with N-chloro amides or imides in a Banbury mixer [12].

The chlorinated polymer can be modified by incorporating unsaturated acids (or anhydrides), unsaturated silanes, or transition-metal coupling agents [e.g. 13,14].

MATERIALS AND TESTING

The bonding properties of polypropylene were studied by measuring the shear strength of single-lap joints between polypropylene and aluminium, bonded with epoxy or polyurethane adhesive.

Two grades of polypropylene were used: X5S-770, a black, rubber-toughened, mineral-filled polymer; and GWM-213, an ethylene-propylene copolymer. Table 1 lists some of their properties.

TABLE 1
Grades of polypropylene used for bonding [15,16]

	X5S-770	GWM-213
Description	Black, rubber-toughened polypropylene (filled)	Ethylene-propylene copolymer (unfilled)
Density (g/cm^3)	0.930	0.905
MFI $(g/10min)$	5	2
Hardness (R)	40	75
Tensile Yield Stress (MPa)	15	22.8

Three adhesives were used in this study. Most of the work used 3M 9323 toughened epoxy. Some studies were also carried out using 3M 3532 polyurethane and Evode DP70-0127 experimental epoxy.

TABLE 2
Commercial adhesives used for bonding polypropylene.

	9323	3532	0127
Description	2-part epoxy	2-part polyurethane	2-part epoxy
Shear Strength* (MPa)	39	7-8	28
Work Life (mins)	120	9	20
Service temps (°C)	-50,80	-50,80	not given
Cure Regime	2h,65°C	48h,20°C	$\frac{1}{2}$ h,85°C

* typical RT values

Various primers from two suppliers were used to treat the polypropylene before bonding. The data sheets provided by the manufacturers describe the primers as being chlorinated polyolefins (or polypropylene). The chlorine contents and molecular weights of the primers are quoted, although the latter property was determined for some primers by GPC analysis.

TABLE 3
Primers used for treating polypropylene before bonding.

Manufacturer	Grade	Cl content (%)	m.w.
Toyo Kasei (Hardlen)	14-LLB	27	45 000
	14-ML	24	45 000
	13-LP	26	120 000
	17-LP	35	115 000
Eastman Chemicals	343-1	18	40 000
	343-3	30	44 000
	515-2	27	31 000

Hardlen data from ref [17].
Eastman data from ref [18] except m.w. figures which were determined by GPC analysis.

EXPERIMENTAL

Preparation of lap joints.

The polypropylene was supplied as injection-moulded plaques 3mm thick x 158mm square, which were bandsawed into coupons 20mm x 60mm. These were then surface-treated and bonded to identical coupons of gritblasted aluminium (1.6mm thick) using a structural adhesive. All the joints had an overlap area of 10mm x 20mm.

The joints were held together by foldback clips while curing, and the glueline thickness was regulated by adding 1% by weight of 250μ glass ballotini to the adhesive during mixing.

After curing the adhesive, the joints were left for a few days at room temperature and were then either tested in shear or transferred, unstressed, to a humid-ageing cabinet (42°/48°/42°C, 95% RH).

Testing of lap joints.

Immediately before testing the joints, the polypropylene coupons were reinforced by backing them with coupons of mild steel (1.2mm thick) using a cold-cure epoxy. This prevented them deforming under load and enabled bond strengths to be measured easily, even when they were near or above the tensile yield stress of the polymer. (The polypropylene coupons were wiped with primer solution before backing them.)

The joints were loaded into an Instron Universal Testing Machine using packing pieces, and were then sheared at a crosshead speed of 2mm/min. Force-displacement plots were recorded and the failure loads of the joints were determined from these.

The shear strength values reported here are the average from batches of six identical joints. The small sample standard deviations have been omitted but were very low for SiC-abraded specimens.

RESULTS

The following tables give average failure loads (kN) of single-lap joints between aluminium and steel backed polypropylene.

Table 4 gives shear strength for joints prepared using 'standard' treatment methods.

Table 5 gives shear strengths for joints prepared using primers on smooth and roughened polypropylene, and in Table 6, the results obtained by 'through-primer abrasion' are compared with those from a set of control joints in which the polypropylene coupons were abraded dry before priming.

TABLE 4
Lap-shear strengths (kN) : standard pretreatments.

	X5S-770		GWM-213	
	Smooth	G'blasted	Smooth	G'blasted
Corona discharge:[a]				
2mJ/mm^2	0.94	0.94	0.66	0.78
10mJ/mm^2	1.20	1.06	0.80	0.95
18mJ/mm^2	1.26	1.23	1.43	1.23
36mJ/mm^2	1.27	1.14	1.42	1.42
Chromic acid;				
Normal conc.[b]	1.66	1.62	2.18	2.05
Diluted 3x[c]	0.31	0.74	0.00	0.33
Plasma[d]	0.4	0.4	0.9	0.7
Flame[e]	(0.2)			
Solvent wipe[f]	(0.1)			
Gritblast[g]	(0.6)			

a : energy refers to discharge process, not polymer surface
b : 15min, 40°C, $K_2Cr_2O_7$: H_2O : H_2SO_4 (7:12:150)
c : 15min, 70°C
d : carried out at Polaron; 10min, 30W
e : carried out at Aerogen; 10:1 air:gas, 3cm/sec
f : trichlorethane
g : 80/90 mesh alumina grit, 100psi

TABLE 5
Lap-shear strengths (kN) : chlorinated primers

	X5S-770		GWM-213	
	Smooth	G'blasted	Smooth	G'blasted
14-ML[a]	1.48	1.42	1.60	1.47
343-1[b]	1.40	1.35	0.94	1.25
343-3[c]	1.20	1.19	0.83	1.18
343-3[d]	1.23	1.21	1.13	1.20
515-2[e]	0.68	0.78	0.89	1.35

a : 5-10% in toluene
b : 13%
c : 17%
d : 3 %
e : unknown concentration. Later experiments using this material gave very low shear strengths; the material is actually described as a stir-in additive rather than a primer.

THROUGH-PRIMER ABRASION

To examine the effects of through-primer abrasion on joint strength, polypropylene coupons were roughened using silicon carbide paper, applying normal hand-pressure, across the width of the coupons. This process was carried out (a) on dry polymer coupons which were then dipped in a primer solution, and (b) on polymer coupons already wet with the primer solution. These through-abraded specimens were also dipped to restore the primer layer thickness.

One set of results examined the effect of through-primer abrading without restoring the primer film by dipping.

TABLE 6
Lap-shear Strengths (kN) : SiC-abraded Polypropylene
Pre-Primer Abrasion and Through-Primer Abrasion

(i) using Hardlen 14-ML primer

	X5S-770		GWM-213	
SiC grade	PPA	TPA	PPA	TPA
120 mesh	1.15	1.29	1.42	1.92
180 mesh	1.24	1.38	1.54	2.03
240 mesh	1.19	1.32	1.43	1.83
400 mesh	1.14	1.41	1.34	1.73

(ii) using Eastman 343-3 primer

	X5S-770		GWM-213	
240 mesh	1.47	1.51	1.62	1.98

(iii) using Hardlen 13-LP and 17-LP primers

	Smooth	PPA[a]	TPA[b]	TPA-R[c]
13-LP (X5S)	0.89	1.06	1.27	1.19
13-LP (GWM)	0.71	1.47	1.80	1.76
17-LP (X5S)	1.29	1.24	1.28	1.27
17-LP (GWM)	1.29	1.53	1.87	1.85

a : 180 mesh SiC
b : 220 mesh SiC, primer layer not restored
c : 220 mesh SiC, primer restored by dipping

RESULTS FROM OTHER ADHESIVES

The following results are average shear-strengths (kN) of lap joints prepared using various surface treatments and bonded with either (a) 3M 3532 polyurethane or (b) DP70-0127 modified epoxy. They are included here to show that even when used with adhesives other than standard epoxy resin, these primers give strong bonds to polypropylene.

(a) 3532 polyurethane (3M)

	X5S-770		GWM-213	
	Smooth	G'blasted	Smooth	G'blasted
Chromic Acid (15min 40°C):	1.44	1.43	1.86	2.13
Corona Disch. (15mJ/mm^2):	1.36	1.20	1.74	1.40
Hardlen 14-ML primer dip:	1.23	1.22	1.63	1.67

(b) DP70-0127 epoxy (Evode)

(i) PP surface treated by pre- and through-primer abrasion using Fibral Nylon pads (grade 4370)

primer	X5S-770		GWM-213	
	PPA	TPA	PPA	TPA
14-ML	0.78	1.18	0.67	1.76
343-3	0.74	0.99	0.77	1.25

(ii) PP surface primed with 14-LLB. Bonded joints humid-aged for 500h before testing. Average joint strengths (kN) : 0.97 (smooth), 1.26 (gritblasted).

Typical unaged joint strength for this system : 1.35kN.

OTHER WORK

Several surface treatment methods were found to be ineffective or unreliable when used on polypropylene, giving low bond strengths with epoxy adhesives.
These included:

(a) Satreat : trichloroisocyanuric acid was applied from solution to polypropylene in an attempt to directly chlorinate the surface. This gave a slight increase in bond strength to rubber-toughened PP (X5S-770) but was otherwise useless.

(b) Fibral : nylon abrasive pads (manufactured by Fibral) were used to roughen the surface of PP before applying a primer. These tests were inconclusive : some 'gritty' nylon pads gave reasonably strong joints, while others gave weaker joints than smooth-primed controls.

(c) Primers : two grades of primer gave notably weaker joints than all the others.
These were:

(i) Eastman 515-2. Although this material gave some good results (Table 5) it is described as a stir-in additive rather than a primer. Later work with this primer produced very weak joints.

(ii) Hardlen 14-LLB-NT. This is a low-m.w. primer which was used on smooth and abraded PP, giving low joint strengths in all cases.

DISCUSSION

The results in Table 6 show (a) that abrasion with SiC paper, followed by priming, gives bonds which are consistently strong; and (b) that abrading the surface of polypropylene while wet with a toluene-borne primer significantly increases the bond strength obtained.

This strength increase may be due to mechanochemical grafting, as described by Lerchenthal [19,20,21] although Kinloch, reviewing that author's work, claims that the improved bond strength could result from forced wetting during abrasion [22].

Mechanochemical grafting required the production of free radicals, the presence of which could be verified by e.s.r. techniques [23,24]; Lerchenthal used a coloured radical-trapping agent instead.

An alternative approach has been suggested [25] whereby the through-primer abrasion is carried out using a primer solution doped with a free radical inhibitor, (e.g. topanol). If the expected strength increase was not observed, then this would imply that free radicals were involved and that the strength increases are not simply the result of mechanical roughening.

Although abrasion under a toluene solution may produce a rougher surface than dry abrasion, this cannot explain completely the observed strength increase; otherwise dry abraded specimens (120 mesh SiC) would be stronger than through-primer abraded specimens (400 mesh SiC).

While the through-primer abrasion process is not new, its application to commercial primers such as chlorinated polypropylene does not appear in the literature. The closest allied process known to the authors is described in two patents [26,27] and involves irradiating a primed surface with UV light so as to generate radicals within the primer and at the surface of the polymer substrate.

According to the papers published by Fujimoto et al [11,28], the primer-substrate bonding involves both diffusion and epitaxial crystallisation, and so the resulting bond strength will depend on the crystallinity (and thermal and mechanical history) of the polypropylene surface.

Further research should therefore be carried out to examine these primer systems, including a detailed analysis of every structural feature (chemical and physical) of the substrate surface.

REFERENCES

1. BRIGGS, D.
   Surface Treatments for Polyolefins.
   In: Surface Analysis and Pretreatment of Metals and Plastics, Ed. D.M. Brewis.

2. WU, S.
   Polymer Interface and Adhesion, ch.9.
   Marcel Dekker Inc. NY (1982).

3. GARBASSI, F., OCCHIELLO, E. and POLATO, F.
   Surface Effect of Flame Treatments on Polypropylene,
   J. Mater Sci., v.22, pp.207-212 (1987).

4. GARBASSI, F., OCCHIELLO, E., POLATO, F. and BROWN, A.
   Surface Effect of Flame Treatments on Polypropylene,
   pt.2. SIMS (FABMS) and FTIR-PAS studies.
   J. Mater Sci., v.22, pp.1450-1456 (1987.

5. BLAIS, P., CARLSSON, D.J., CSULLOG, G.W. and WILES, D.M.
   The Chromic Acid Etching of Polyolefin Surfaces, and Adhesive Bonding.
   J. Coll. Interface Sci., v.47, n.3, pp.636-649 (June 1974).

6. BRIGGS, D., ZICHY, V.J.I., BREWIS, D.M., COMYN, J., DAHM, R.H., GREEN, M.A. and KONIECZKO, M.B.
   X-Ray Photoelectron Spectroscopy Studies of Polymer Surfaces
   4 - Further Studies of the Chromic Acid Etching of Low Density Polyethylene.
   Surf. Interface Anal., v.2, n.3, pp.107-114 (1980).

7. DA COSTA, R.A., GONCALVES, M.d-C, DE OLIVEIRA, M.G., RUBIRA, A.F. and GALEMBECK, F.
    Polyethylene Adhesion : Pretreatment with Potassium Permanganate.
    J. Appli. Polym. Sci., v.37, pp.3105-3117 (1989).

8. CARLSSON, D.J. and WILES, D.M.
    Surface Studies by Attenuated Total Reflection Spectroscopy.
    I. Corona Treatment of Polypropylene.
    Canadian J. Chem., v.48, pp.2397-2406 (1970).

9. STROBEL, M., DUNATOV, C., STROBEL, J.M., LYONS, C.S., PERRON, S.J. and MORGEN, M.C.
    Low-Molecular-Weight Materials on Corona-treated Polypropylene,
    J. Adhesion Sci. Technology, v.3, n.5, pp.321-335 (1989).

10. BEEVERS, A., THERNOE, J. and NJEGIC, A.
    Corona Discharge Pretreatment of Surfaces for Bonded Joints.
    IMechE., conf. proc., Bristol 1986.

11. FUJIMOTO, F.
    Properties and Applications of Chlorinated Polypropylene.
    Paint & Resin, pp.36-40, (Feb 1986).

12. SAGANE, T. and NAGANO, R.
    Laminated structure having post chlorinated adhesive olefin resin layer.
    European Patent 0 149 356 A2, 24/07/85.

13. ITO, T., MOGAMI, M. and KOBAYASHI, K.
    Coating Resin Composition.
    European Patent 0 226 387 A2, 24/06/87.

14. MAYUMI, J. and MARUTA, R.
    Primer Composition for Olefin Resin.
    European Patent 0 193 126 A1, 03/09/86

15. ANON.
    ICI 'Propathene' *Data Summary Sheet*, PP.34 (6th Ed.)
    Polypropylene - Injection Moulding Grades (1986).

16. ANON.
    ICI 'Procom' *Data Summary Sheet*, PM.63 (2nd Ed.)
    Modified, filled and reinforced polypropylenes for infection moulding and extrusion (1986).

17. ANON.
    'Hardlen' *Information Booklet*, Toyo Kasei Kogyo Co., Ltd. (c.1986?).

18. MILLER, S. and MIDDLETON, K.
    Coatings for Polypropylene and Other Selected Substrates,
    Coatings for Plastics Symp., Paint Research Association (Harrogate, May 1986).

19. BRENMAN, M. and LERCHENTHAL, C.H.
    Mechanochemistry and Adhesion : Improved Strength of
    Polymeric Joints.
    J. Applied Polym. Sci., Appl. Polym. Symp.35, pp.537-543
    (1979).

20. LERCHENTHAL, C.H. and BRENMAN, M.
    Increase of Adhesive Bond Strength through the
    Mechanochemical Creation of Free Radicals.
    Polym. Eng. Sci., v.16, n.11, p.747 & p.760 (1976).

21. LERCHENTHAL, C.H., BRENMAN, M. and YITSHAQ, N.
    J. Polym. Sci., Polym. Chem. Ed., v.13, p.737 (1975).

22. KINLOCH, A.J.
    Adhesion and Adhesives Science and Technology, p.106,
    Chapman and Hall (1987).

23. SOHMA, J.
    Mechanochmistry of Polymers
    Prog. Polym. Sci., v.14, pp.451-596 (1989).

24. SIMIONESCU, C. and OPREA, C.V.
    Mechanochemical Synthesis
    Russ. Chem. Rev., v.57, n.3, p.283 (1988).

25. DR. B.D. LUDBROOK
    Private Communication.

26. BRAGOLE, R.A.
    Method for Bonding Adhesives to Polyolefin Surfaces and
    the Laminate formed.
    World Patent 88/05346, 28/07/88.

27. BASEDEN, G.A.
    Process for improving the adhesion of paint to polyolefin
    surfaces.
    US Patent 4 303 697, 01/12/81.

28. ASHIHARA, T., OHNISHI, A., OKANO, Y. and FUJIMOTO, F.
    Adhesive Properties of Modified Chlorinated
    Polypropylene,
    Conf. Proc., International Meeting, The Adhesion Society
    (1987).

5

BONDING OF POLYOLEFINS WITH CYANOACRYLATE ADHESIVES

Patrick F. McDonnell
Research and Development Department,
Loctite (Ireland) Limited,
Dublin,
Ireland.

INTRODUCTION

Since their commercialisation in the late fifties, cyanoacrylate adhesives have become well established in both the industrial and consumer markets [1]. Bonds with high mechanical strength and good durability can be achieved in seconds on a wide range of materials. A minimum of surface preparation is required in most cases. The anionic curing mechanism is initiated by traces of moisture on the surface or by the intrinsic reactivity of the substrate. Certain neutral or acidic materials, such as wood, paper or ceramics, can also be bonded with cyanoacrylate adhesives containing an ion complexing agent in the formulation [2-5].

While almost universal in their application, cyanoacrylate adhesives have heretofore had the major limitations of being ineffective as bonding agents for low energy polymeric materials typified by polyethylene and polypropylene. These plastics, together with derivatives such as certain thermoplastic rubbers, continue to achieve high growth due to their combination of low cost and material properties. Polyolefins comprise about 40% of the plastics used in industry and there has been a need for a facile means of joining these materials to themselves and to other substrates

Various treatments exist to improve the bondability of polyolefins. In general, the surface is modified to increase the surface energy by introducing or forming polar groups, by methods such as flame treatment, corona discharge, irradiation, oxidation with such as chromates or sulphuric acid and plasma treatment [6]. Special conditions, equipment and techniques, frequently expensive and/or hazardous, are needed for these processes. These treatments are often ineffective for joints involving cyanoacrylate adhesives since acidic residues generated by the treatments inhibit the normal hardening.

There is, therefore, a need for a surface treatment for polyolefins which is convenient to use and would exploit the full benefits of bonding with cyanoacrylate adhesives. This paper is concerned with various methods which achieve this objective.

POLYOLEFIN PRIMERS

There have recently been a number of disclosures in the patent literature describing the use of primers for bonding polyolefins with cyanoacrylate adhesives. Some of these are reviewed below. The primers consist of dilute solutions of various active materials in organic solvents. A coating of the primer is applied to the polyolefin surface to be bonded and allowed to dry. A cyanoacrylate adhesive is then applied to complete the bond. In general, the primers function at normal room temperatures with a minimum of process complexity.

The primer coating can be almost monomolecular in thickness with concentrations in the range 0.005 to 0.025 g/m^3 reported [7]. The corresponding quantity of cyanoacrylate adhesive would be approximately 10^5 times this level. Primers can be formulated in carrier solvents so that approximately equal volumes of primer and adhesive are laid down during a bonding operation.

Primer Compositions/Patent Review

A brief review of polyolefin primers disclosed in recent patent publications shows the diversity of active materials which are claimed to be effective.

Primers based on a combination of an organometallic compound, a
resin and a fluorine containing compound are claimed to bond polyolefins
with cyanoacrylates and also with other adhesives. The organometallic
compound can be manganic acetylacetonate [8].

Other primer compositions based on organometallic compounds have
also been reported. These include an extensive list of formulations
involving mixtures of, e.g. 1,3-dicarbonyl complexes and chlorinated
polyolefin resins dissolved in toluene [9].

Guanidine, biguanide or derivatives in solvents such as freons,
acetone or ethanol are reported as primers [10]. EVA copolymers and
chlorinated polyethylene are also mentioned in the compositions.

Primers based on organic phosphine compounds such as phenyl
phosphine and bis-diphenyl phosphinomethane are reported [11].
Solutions of bis (2,4-di-tert-butyl phenyl pentaerythritol diphosphite)
in toluene are also disclosed [12].

A number of primer compositions based on solutions of various amines
have also been reported, of which the following are examples. Primers
based on an amine in combination with an aldehyde compound such as
dimethylaminobenzaldehyde [13]. Tertiary amine compounds such as
2-dimethylaminoethanol or 4-dimethylamino-1-butanol [14]. Aminosilane
compounds such as N-beta (aminoethyl) gamma-aminopropyl-trimethoxysilane
[15]. Primers based on polyamine compounds having at least two amino
groups, e.g. a 0.25% w/w solution of triethyltetraamine in Freon 113
[16]. Primers based on solutions of 4-vinylpyridine or lutidine in
various solvents [17]. Compositions involving heterocyclic amines
such as 1,5,7-triazabicyclo [4.4.0] dec-5-ene have also been disclosed
[18].

A further disclosure describes a primer composition containing a
chlorinated or carboxylated polyolefin and a cross-linking agent such as
an amine, amidoamine, polyamide, etc. [19].

Bonding Performance
The remainder of this paper will concentrate on the performance of two
recently commercialised primers. Both are manufactured by Loctite
Corporation and have the commercial names of Polyolefin Primers 757 [20]
and 770.

Primer 757 is formulated in Freon TA, an azeothrope of acetone and 1,1,2-trichloro 1,2,2-trifluoroethane. This solvent has the benefits of fast drying and non-flammability. Primer 770 contains heptane as solvent. Both primers have the same active ingredients and are low viscosity fluids. Both also contain a U.V. fluorescent agent to aid confirmation that primer has been applied to the bond area.

MEASUREMENT METHODOLOGY

Bonding tests were carried out on various substrates primed with Polyolefin Primers 757 and 770 in combination with various grades of cyanoacrylate adhesives. Unless otherwise stated, a low viscosity, liquid grade (e.g. Loctite 406) was used in the following tests. Polymeric substrates were cut to dimensions of 100 mm x 25 mm x 3 mm, deburred and cleaned with isopropyl alcohol. Metal substrates had dimensions of 100 mm x 25 mm x 1.6 mm and were degreased before use.

Primers 757 or 770 were applied to the test pieces (lapshears) by brush or spray to form a single coating over the area to be bonded. The coatings were allowed to dry under unforced conditions. In general, bonds were prepared within 1 to 5 minutes after the coating had dried. The primers were applied only to low energy materials such as polyolefins or polyacetals and not to more easily bondable materials such as metals and many plastics. It should be noted that priming easy to bond materials was found to decrease bond strength to these substrates.

Joints with an overlap bond area of 160 mm^2 were constructed using a cynaoacrylate adhesive. The primers also accelerated the cure of the adhesive making clamping unnecessary. Bond strengths were determined under tensile shear mode after a 24 hour cure at room temperature. Testing was to ASTM D1002 using a Universal testing machine. The crosshead speed was 2 mm/minute for rigid materials and 100 mm/minute for bonds involving elastomeric substrates.

Bondable Materials

Apart from common polyolefins, other low surface energy materials such as polyacetals and thermoplastic rubbers can be bonded by these methods. Bond strengths in excess of 6 N/mm² can readily be obtained on rigid grades of polyethylene, polypropylene and polyacetal and substrate failures are achievable at lower bond strengths with elastomers (Table 1).

TABLE I

Bond Strengths Achieved on Various Low Energy Materials With
Primer 757 and a Cyanoacrylate Adhesive

Substrate	Tensile Shear Bond Strength (N/mm²)	
	Primer 757	No Primer
Polyethylene (Simona, HWST)	6.0	0
Polypropylene (Simona, DWST)	8.9	1.2
Polyacetal (Hoechst Celanese, Kematal)	2.9	0
Polyacetal (E.1 DuPont, Delrin)	7.7	0
Thermoplastic Rubber (Monsanto, Santoprene, Various Grades)		
101-64	0.7*	0.2
101-73	1.11*	0.3
101-87	2.3*	0
103-40	3.2*	0
103-50	4.1*	0
203-50	6.1*	0
202-87	1.33*	0
PTFE	3.0	0
Polymethylpentene (I.C.I., TPX)	6.3	0

*Substrate Failure.

Choice Of Cyanoacrylate Adhesives

The primers tested above are effective only when used with cyanoacrylate
adhesives. These adhesives are now available in several grades which
vary in such properties as viscosity, cure speed, monomer basis and
toughness. Within the family of cyanoacrylate adhesives, the low
viscosity liquid grades gave the highest bond strengths. The more
specialized toughened and gel-form grades of ethyl cyanoacrylate gave
relatively lower bond strengths (Table 2).

TABLE 2

Bond Strengths Achieved on Polyethylene With Primer 757 and
Various Types of Cyanoacrylate Adhesives

Cyanoacrylate Adhesive Type	Tensile Shear Bond Strength (N/mm^2)	
	Primer 757	No Primer
Low Viscosity, Fast Curing	6.5	0
High Viscosity	4.6	0
Medium Viscosity, Fast Curing	6.3	0
Toughened	2.2	0
Thixotropic	4.1	0
Low Odour, Alkoxyester	1.3	0

Bonding Mixed Substrates

In practice, many joints involving polyolefins will have a second
substrate which is easily bonded by ethyl cyanoacrylate based
adhesives. In these situations, maximum bond strengths were obtained
when the primer was applied only to the polyolefin and the cyanoacrylate
adhesive to the more easily bonded materials. Mixed substrate bonds
prepared in this way between polyolefins and steel, aluminium
polycarbonate and polymethylmethacrylate gave substantial bond strengths
in comparison to bond strengths for unprimed controls (Table 3).

TABLE 3

Bonding Mixed Substrates, One of Which is a Low Energy Material,
With Primer 757 and a Cyanoacrylate Adhesive

Bond		Tensile Shear Bond Strength (N/mm^2)	
		Primer 757	No Primer
Polyethylene (Simona, HWST)	- Mild Steel	4.7	0
"	- Aluminium	3.1	0
"	- Polycarbonate	3.1	0
"	- PMMA	2.9	0
Polypropylene (Simona, DWST)	- Mild Steel	5.1	0
"	- Aluminium	2.1	0
"	- Polycarbonate	6.4	0
" •	- PMMA	3.7	0
Thermoplastic Rubber (Monsanto 101-64)	- Mild Steel	0.43	0
"	- Aluminium	0.43	0
"	- Polycarbonate	0.50	0
"	- PMMA	0.43	0

Bonding Filled Grades Of Polypropylene

Many commercial grades of polypropylene contain fillers or additives
such as talcs, stabilizers and flame retardants. The bondability of
polypropylene with cyanoacrylate adhesives was found to be greatly
influenced by the type of additive. While untreated PP containing
additives such as U.V. stabilizers, release agents, talc and heat
stabilizers were unbondable with cyanoacrylate adhesives, other
additives such as elastomers and anti-static agents resulted in
relatively high bond strengths. Regardless of the type of filler, the
use of primers was found to give high bond strengths with cyanoacrylate
adhesives (Table 4).

TABLE 4

Bonding Various Grades of Filled Polypropylene (ex Hoechst)
With Primer 770 and a Cyanoacrylate Adhesive

Filler Type	Tensile Shear Bond Strength (N/mm^2)	
	Primer 770	No Primer
Talc	7.5*	0
Release Agent	7.1*	0
U.V. Stabilizer	6.9*	0
Heat Stabilizer	7.6*	0
Flame Retardant	6.0*	1.9
Talc/Wax	7.6	2.8
High Crystallinity	6.7	3.2
Elastomer	6.7	3.4
Anti-static (Food Grade)	7.7	3.8
Anti-static (Non-Food)	8.0	4.5
"Powder"	6.0	5.2

*Substrate Failure.

Adhesive Cure Speed With Primers

As mentioned earlier, Primers 757 and 770 also accelerate the cure of cyanoacrylate adhesives with very rapid development of ultimate bond strength. Fixture or handling strength was achieved on polypropylene in less than 30 seconds and 80% of ultimate bond strength in one hour. All bonds were prepared and cured at normal room temperature (Table 5).

TABLE 5

Rate of Bond Strength Development for Polypropylene With
Primer 757 and Bonded with Cyanoacrylate Adhesive

Time	Tensile Shear Bond Strength (N/mm^2)
30 Seconds	1.7
360 "	3.1
1 Hour	7.2
8 "	9.0
100 "	9.8

It should be noted that many amine type compounds applied from solution will rapidly cure cyanoacrylate adhesives on polypropylene. However, the resulting polymer has little or no adhesive effect. Specific compounds also promote adhesion in the way described in this paper.

Primer On-Part Life

The active on-part life of such primers, which is a very practical issue in their identified use, was also examined. Polypropylene test pieces treated with Primer 757 were found to give the maximum bond strength if the cyanoacrylate adhesive was applied and the bond completed within an hour of treatment. If bonding after treatment was delayed, a gradual reduction in bond strength was observed with a significant fall-off after 8 hours (Table 6). Re-treatment with fresh primer was found to restore the activity.

TABLE 6

On Part Life of Primer 757 and Subsequent Bond Strength
With Cyanoacrylate Adhesive on Polypropylene

On-Part Period*	Tensile Shear Bond Strength (N/mm^2)
10 Seconds	8.5
30 "	9.0
60 "	8.8
300 "	8.7
8 Hours	5.9
24 "	2.7

*Primer 757 applied to polypropylene surface and left exposed to
air at R.T. for the on-part period.

CONCLUSIONS

1. Primers have been developed which act as effective adhesion promoters in the bonding of polyolefins and other low surface energy materials with cyanoacrylate adhesives.

2. The primers consist of dilute solutions of specific compounds in organic solvents.

3. Two primers were evaluated for various bonding characteristics and high bond strengths together with very rapid adhesive cures were observed.

4. The primers are only effective with cyanoacrylate type adhesives and the detailed mechanism by which they operate remains to be established.

REFERENCES

1. K. L. Shanta, S. Thennarasu and N. Krishnamurti, J. Adhesion Sci. Technol. Vol. 3. No. 4. pp. 237-260 (1989).

2. Loctite Corporation. U.S. Patent 4,636,539 (1987).

3. Toa Gosei Chem. Ind. Co. U.S. Patent 4,837,260 (1989).

4. Toa Gosei Chem. Ind. Co. U.S. Patent 4,170,585 (1979).

5. Taoka Chem. Co. Eur. Patent 0063037 (1982).

6. L. E. Rantz, Adhesives Age, May 15, 1987, pp. 10-16 (1987).

7. Toa Gosei Chem. Ind. Co. Ger. Patent 3601518 (1988).

8. Toa Gosei Chem. Ind. Co. U.S. Patent 4,822,426.

9. Toa Gosei Chem. Ind. Co. Eur. Patent 129069 (1984).

10. Toa Gosei Chem. Ind. Co. Japan Patent 64001784 (1989).

11. Toa Gosei Chem. Ind. Co. Japan Patent 63162781 (1988).

12. Toa Gosei Chem. Ind. Co. Japan Patent 63010796 (1988)

13. Three Bond Co. Eur. Patent 271675 (1988).

14. Three Bond Co. Japan Patent 63128090 (1988).

15. Dainichi Nippon Cables. Japan Patent 63056584 (1988).

16. Mitsui Petrochemical Ind. Japan Patent 63051489 (1988).

17. Alpha Giken. Japan Patent 62018486 (1987).

18. Loctite Corporation. U.S. Patent 4,869,772 (1989).

19. A. Vanmeesche. Eur. Patent 187171 (1986).

20. P. McDonnell, D. Melody, Adhecom 89 (1989).

6

POLYURETHANES AND ISOCYANATES CONTAINING HYDROPHILIC GROUPS
AS POTENTIAL COMPONENTS OF WATER-BORNE ADHESIVES

Janusz Kozakiewicz
Industrial Chemistry Research Institute
01-793 Warsaw Poland

1. INTRODUCTION

Polyurethanes and isocyanates are widely used in the adhesives industry not only as substantial binders, but also as various additives to adhesives based on the other polymers. Introducing hydrophilicity to polyurethanes or isocyanates creates new interesting possibilities of their application in water-borne adhesives which, being non-toxic and non-flammable, are probably the most developing group in the adhesives industry.

Depending on the proportion of hydrophilic groups introduced into a structure of the polymer that is usually hydrophobic, two types of polyurethanes with different grades of hydrophilicity can be considered:

- water-dispersable - up to ca 10% hydrophillic groups
  products
- water-soluble products - ca 50%-70% hydrophillic groups

Aqueous Dispersions of Polyurethanes (DPU)

Aqueous dispersions of polyurethanes (DPU) seem to be at present one of the most developing branches of polyurethane chemistry. This is because of strict limitations concerning the use of solvent-based systems in the industry. A paper dealing with this problem was presented at this Adhesion Conference a few years ago [1].

Studies on DPU must have been really very intense since more than 1000 patents were recorded until 1980 [2] and even now ca40 patents appear every year. DPU manufactured nowadays are usually ca40% aqueous dispersions (the dispersing phase may also be a mixture of water and an organic solvent that is miscible with water) of polyurethanes or polyurethane-ureas with hydrophilic (ionic or non-ionic) moities. The content of hydrophilic groups is such that the polymer is able to self-disperse in water. Due to this "internal" dispersing system, it is not necessary to use emulgators and therefore, supposing the formulation and technology is correct, films of dried DPU show better water resistance than their analogues obtained from the other dispersions (eg acrylic).

Because of high reactivity of the isocyanate group, several synthetic routes leading to dispersable polyurethanes with hydrophilic moities can be considered, but what seems to be the most important is not the chemistry but the technology of transferring such partially hydrophilic polymer to aqueous dispersion. Presumably, three technogical processes are at present used in industrial practice [3,4,5]:

- "acetone" process   - Fig 1
- "prepolymer-ionomer" process   - Fig 2
- "melt" process   - Fig 3

The main difference between these processes lies in the structure of the products. In "prepolymer-ionomer" and "melt" processes, the resulting polyurethane (or

Stage I. : Synthesis of NCO-terminated urethane prepolymer.

HO~~~~OH + OCN━━━━NCO ------> OCN-ᐧᐧᐧᐧᐧᐧ-NCO
Polyol   Diisocyanate   Urethane prepolymer

Stage II. : Diluting the prepolymer with acetone.

Stage III, IV and V : Introducing hydrophilic moieties or their
                      precursors to the prepolymer, dispersing
                      the resulting polyurethane in water and
                      distillation of acetone.

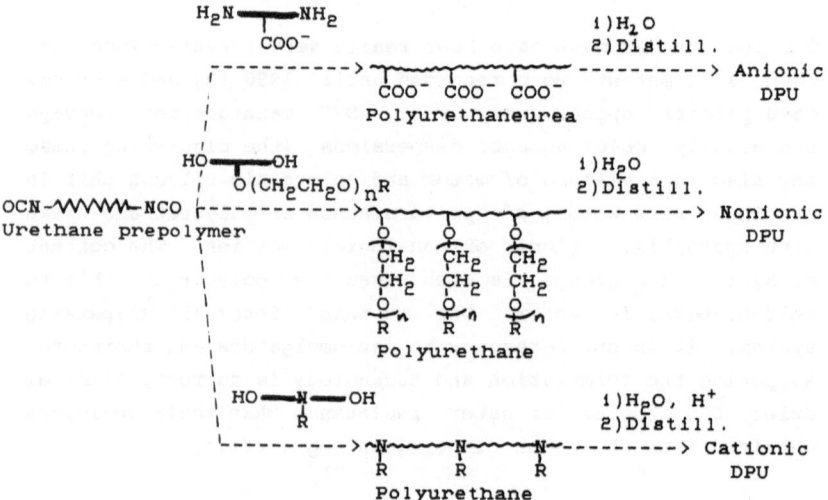

Fig 1   "Acetone" process of DPU synthesis. Three possible
        routes leading to anionic, cationic and non-ionic
        dispersions are presented. In practice, usually only
        anionic, cationic and mixed anionic/non-ionic or
        cationic/non-ionic products are offered.

        ~~~~ - polyol chain
        ∧∧∧∧ - urethane prepolymer chain
        ━━━━ - fragments of other compounds' structures
        ~~~~ - polyurethane or polyurethaneurea chain

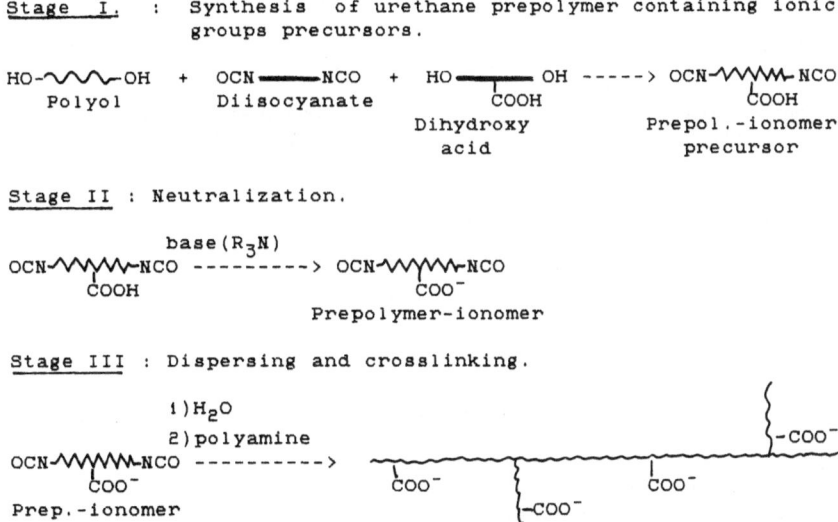

Stage I. : Synthesis of urethane prepolymer containing ionic
           groups precursors.

HO-〜〜-OH + OCN━━NCO + HO━━OH -----> OCN-〜〜〜-NCO
   Polyol   Diisocyanate   COOH   COOH
                                  Dihydroxy   Prepol.-ionomer
                                    acid   precursor

Stage II : Neutralization.

                 base(R₃N)
OCN-〜〜〜-NCO ---------> OCN-〜〜〜-NCO
     COOH   COO⁻
                         Prepolymer-ionomer

Stage III : Dispersing and crosslinking.

            1)H₂O
            2)polyamine
OCN-〜〜〜-NCO ---------->
     COO⁻
Prep.-ionomer

                              Croslinked anionic DPU

Fig 2   "Prepolymer-ionomer" process of DPU synthesis. A
        route leading to anionic dispersion of crosslinked
        polyurethaneurea is presented.

         〜〜〜 - polyol chain
         〜〜〜 - urethane prepolymer chain
         ━━━━ - fragments of other compounds' structures
         〜〜〜 - polyurethane or polyurethaneurea chain

Stage I and II : As in prepolymer-ionomer process.

Stage III : Amidation.

$$\underset{\overset{\displaystyle |}{SO_3^-}}{ONC\text{-}\mathsf{WWW}\text{-}NCO} \xrightarrow{\quad NH_3 \ (urea) \quad} \underset{\overset{\displaystyle |}{SO_3^-}}{H_2N\text{-}CO\text{-}NH\text{-}\mathsf{WWW}\text{-}NH\text{-}CO\text{-}NH_2}$$

Stage IV : Methylolation and dispersing.

$$\underset{\overset{\displaystyle |}{SO_3^-}}{H_2N\text{-}CO\text{-}NH\text{-}\mathsf{WWW}\text{-}NH\text{-}CO\text{-}NH_2}$$

$$\downarrow \ HCHO/H_2O$$

$$\underset{\overset{\displaystyle |}{SO_3^-}}{HO\text{-}CH_2\text{-}NH\text{-}CO\text{-}NH\mathsf{WWW}\text{-}NH\text{-}CO\text{-}NH\text{-}CH_2\text{-}OH}$$

Stage V : Polycondensation.

$$\underset{\overset{\displaystyle |}{SO_3^-}}{HO\text{-}CH_2\text{-}NH\text{-}CO\text{-}NH\mathsf{WWW}\text{-}NH\text{-}CO\text{-}NH\text{-}CH_2\text{-}OH}$$

$$\downarrow \ \Delta t$$

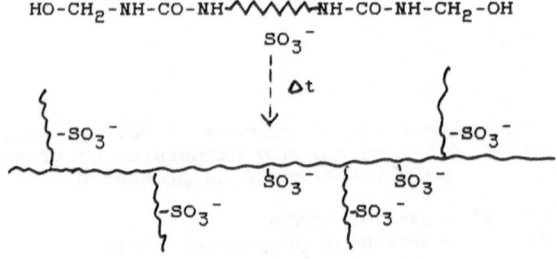

Crosslinked anionic DPU

Fig 3 **"Melt" process of DPU synthesis. A route leading to anionic dispersion of crosslinked polyurethaneurea.**

      WWW - urethane prepolymer chain
      $\sim\!\sim\!\sim$ - polyrethaneurea chain

polyurethane-urea) is usually crosslinked while the one obtained in the "acetone" process is not. This is of great importance since both mechanical strength and water-resistance of the film obtained from DPU would strictly depend on a degree of crosslinking.

However, several crosslinking agents for DPU have been developed and applied in practice. Hence, the "acetone" process, having some advantages over the other two, is still in use, especially if application of the DPU in adhesives is concerned. Polyaziridines, dispersable isocyanates and melamine resins are usually recommended as crosslinking agents for DPU [6,7,8]. The idea of reactions proceeding through crosslinking of DPU with these reagents is presented in Fig 4 a, b and c.

Polyaziridines would have been the most useful crosslinkers for adhesives based on DPU since crosslinking proceeds even at room temperature, but they are quite toxic so the other two are applied more frequently in practice. For those two, a room temperature of 60-80 $^\circ$ C is sometimes enough to start the reaction with polyurethane, but higher temperatures (120-160°C) are always preferred. Of course - the higher the temperature, the shorter is the curing time.

For DPU containing free amino groups there is another interesting possibility for crosslinking - the use of water-borne epoxy resins [7]. As epoxy resins are very good adhesives themselves, this way of crosslinking can be considered as especially interesting for adhesive formulator; however, there is no literature data available on the use of such systems in adhesive applications.

a) Polyaziridine crosslinking.

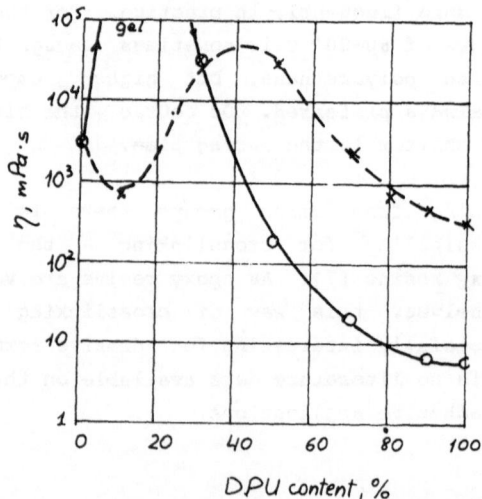

~~~~ + CH₂-CH₂ -----> ~~~~~~ ------> ~~~~~
| \ / |
COO⁻   N⁺   CO-O-CH₂-CH₂NH   CO-N-CH₂-CH₂OH
                 /   \ | |

b) Blocked polyisocyanate crosslinking.

                                    -BLH
~~NH-CO-NH~~ + -NH-CO-BL -----> ~~NH-CO-N~~
Urea bond in DPU |
                                           CO-NH-

c) Melamine resin crosslinking.

                                    -ROH
~~~NH-CO-NH~~~ + -N-CH₂-OR -----> ~~~~NH-CO-N~~~
Urea bond in DPU | |
                                                   CH₂
                                                    |
                                                   N~

Fig 4 Crosslinking of DPU

~~~ - polyurethane or polyurethane-urea chain
BL - NCO-blocking group

Fig 5 Viscosity vs composition of DPU/EVA dispersion blend

O - DPU was not thickened before blending
X - DPU was thickened with 3% Dicrylan
    Verdicker R (CIBA Geigy) before blending

DPU as Adhesive Intermediates

Because of excellent mechanical properties and water-resistance of films, DPU's are used mainly as coatings and finishings in leather and textile industries [9,10,11,12,13]. However, studies on their application as adhesive intermediates have been also carried out, and eventually some companies (BASF, Bayer, ICI) developed DPU that can be used as substantial binders for water-borne adhesives. A pioneer theoretical approach to the problem of adhesion of partially hydrophilic polyurethanes to various solid substrates was given by Prof Lorenz and his research group at Aachen Polytechnic [14,15]. They studied the adhesion in the bonds of films made of DPU and various thermoplastics (PE, PA,, PVC, PC, PS, PMM) and found that in addition to Van der Waals forces, covalent bonds are also formed in the bonding process and are responsible for adhesion. This interesting study not only explained the adhesion problems in such systems, but also proved the possiblity of using DPU as adhesive intermediates.

The advantages and disadvantages of DPU when used in adhesive applications can be specified, as follows:

Advantages	Disadvantages
- no toxicity and flammability problems	- very low viscosity
- no problems in blending with other dispersions	- high surface tension resulting in poor surface wettability
- excellent mechanical strength and water resistance of films (especially when crosslinked)	- lack of film tack
	- high cost

Because of these disadvantages, it seems clear that DPU should be modified before using them as adhesives. The most common modifications comprise:

- thickening to increase viscosity
- blending with other dispersions to reduce cost and improve certain properties
- blending with tackifying resins

Thickening of DPU is very important since it is entirely impossible to bond porous adherends using DPU "as received" - it would immediately penetrate the adherend surface. However, one must be very careful to choose suitable thickeners because certain additives of this sort may not only increase viscosity, but also lower adhesion or even cause gel-formation. As our own experience has shown, acrylic dispersion thickeners are usually the best choice. For instance the viscosity of anionic DPU produced in "acetone" process increased 50 times (from 6 to 300 mPaxs) after 3% of acrylic dispersion-based thickener (Dicrylan Verdicker R - Ciba Geigy) had been added. Polyurethane thickeners that will be discussed later on in this paper are more suitable for acrylic dispersions than for DPU. Blending DPU with other dispersions or latices [16] can lead to systems of interesting chemical structure, especially when they are crosslinked (eg interpenetrating networks). Sometimes, unusual behaviour can be observed in such blends. In our investigations we found that the maximum viscosity is reached in DPU/EVA or DPU/vinyl-acrylic copolymer dispersion blends at certain ratios of both components in the blend. Results obtained for anionic (alkaline) DPU/EVA blends are presented in Fig 5. The explanation of this phenomenon is not clear because even neutralization of an EVA dispersion that is normally acidic, did not change the behaviour of the system. The other intersting observation was that it did matter whether DPU was added to the other dispersion or vice-versa. In fig 6 this is illustrated for DPU/vinyl-acrylic copolymer dispersion blends.

The possibility of obtaining the blends, not after, but during the preparation of DPU (or synthesis of the other dispersion) is quite interesting. Such technique can result

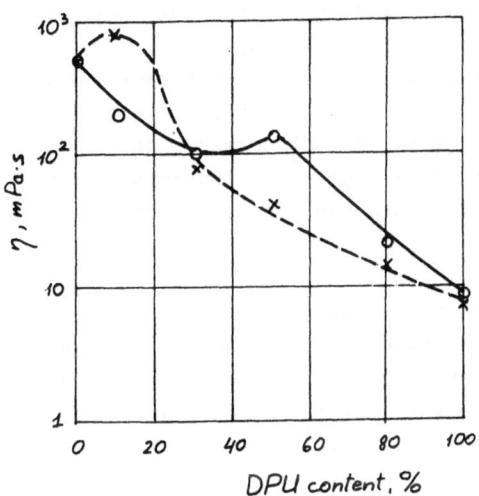

Fig 6   Viscosity vs composition of DPU/acrylic dispersion
blend

O - DPU was added to acrylic dispersion
X - acrylic dispersion was added to DPU

in very interesting properties of the blend since additional
linkages can originate between its components. Though this
idea was presented in several patents (eg in [17]), where
polymerization of acrylic or vinyl monomers carried out in
DPU was claimed, it presumably has not yet been applied in
industrial practice.

We succeeded in obtaining a blend of cationic (acidic) DPU
with PVAc dispersion by adding a solution of polyurethane
containing tertiary nitrogen in the backbone to a commercial
PVAc dispersion (that is strongly acidic) instead of using
diluted acetic acid as a dispersing phase [18]. This blend
has been successfully applied to formulate adhesives used for
laminating PVC decorative films to wood for the furniture
industry.

For certain adhesive applications, especially those where
initial bond strength must be taken into account, tack of the
dispersion which is to be the substantial binder for the
adhesive, is a very important factor. Since DPU's are not
tacky, they must be tackified with low-molecular weight
resins (eg hydrocarbon resins, butylphenolic resins or rosin
and its derivatives). The problem is that only few
commercial resins of this sort are available as aqueous
dispersions, the example can be the set of hydrocarbon resins
offered by Hercules Inc. Others must be added as solutions
in organic solvents, which spoils the the advantage of DPU
over solvent-based polyurethane systems. This should be
considered especially in the footwear adhesive applications
of DPU.

Generally speaking, the literature data concerning
applications of DPU in adhesives is very limited, but in a
number of patents dealing with DPU synthesis the possibility
of their use as adhesive intermediates is almost always
mentioned. Though in one of the papers [2] several bonds
that can be obtained with DPU based adhesives are listed (eg
PVC with PVC, PU foam, metals, PU film, ABC, glass; or PE

with PE and PP with PP), the directions for using DPU in adhesives are still not quite clear.

In several patents it is claimed that DPU can be used in the footwear industry to bond upper parts to soles in shoes instead of solvent-based polyurethane adhesives [19,20,21,22]; in the packaging industry to laminate thermoplastic films [23]; in the textile industry to laminate textile materials [24]. Based on our own results [25] we believe that the adhesives formulated from the blends of DPU with PVAc and EVA dispersions can be useful for bonding PVC decorative films to wood in the furniture industry and also as intermediate adhesive layers in the process of transfer-coating of textiles with PU to make artificial leather. The latter application has been mentioned also in the other paper [10].

Undoubtedly, the possibility of using DPU in the footwear industry has been studied the most extensively, but there are only few papers available presenting the technological problems that can be faced and the practical results that have been obtained [26,27]. The main problems are: drying time that is much longer for DPU based adhesives than for solvent based PU systems and initial bond strength that is too low for the former. The first disadvantage has been overcome by using a stream of hot air (temp ca 65°C) to dry the adhesive films - drying time can be shortened to ca 15 minutes which is a value that can be accepted. To eliminate the second one, formulations containing special tackifiers or blends of DPU with other, more tacky dispersions, (eg acrylic) have been developed.

Though a special single-component DPU has been recently proposed as intermediate for footwear adhesives by some manufacturers (eg Bayer) it is generally necessary to use two-component systems for this purpose to avoid problems with water and heat resistance of the bonds. The possible curatives for DPU have already been discussed earlier in this paper, here some technological aspects of using two-component DPU-based adhesives in the footwear industry will be presented. The main problem is this case is to avoid curing during drying and heat activating of the adhesive film [26,27]. The ideal solution would be to use a heated press instead of heat activation of teh adhesive film before bonding, but this would increase the production cost. Otherwise, curatives must be chosen very carefully and their curing characteristics must be correlated with bonding technology.

If the technology is corect and a high quality adhesive based on DPU is used, the results obtained in bonding shoe soles to uppers can be as good as with solvent based PU systems. The data which have been reported are given in Table 1 [27].

TABLE 1

Materials Bonded	Strength* N/mm	
	Initial (30 min)	Final (24 hrs)
rubber/rubber	5.5	10.5
PVC/PVC	10.5	16.5
PU/nat. leather	7.0	12.0
ruber/nat.leather	11.5	18.5

(*) Usual requirements for (rubber/nat leather) bonds:
   initial strength (5 min) - min 1.5 N/mm
   final strength (24 hrs)   - min 3.0 N/mm

In footwear applications, adhesives based on DPU compete with hot-melts which are solventless too, and offer a much shorter bonding cycle. However, temperature resistance of the bonds with DPU based adhesives is much better (only 40% decrease instrength at $80\,^{\circ}C$) and, so far, they are more promising.

Generally, it can be concluded that the use of DPU's as the substantial components of water-borne adhesives is still developing and, perhaps depending on the possiblities of lowering the cost of commercial DPU, new application areas will appear. At present the main directions of industrial application seem to be footwear and packaging industries.

Water Soluble Polyurethanes (WSPU)

The chemistry of water soluble polyurethanes (WSPU) differs from that of aqueous polyurethane disperions (DPU) [28]. Hydrophilic groups in WSPU are usually non-ionic and are present in much higher quantities in the polymer. So far, the only application of WSPU that can be useful in the field of adhesives is thickening of water borne systems [29,30,31].

Typical a molecule of WSPU presents long hydrophilic segments (usually originally from polyethylene glycol) linked by urethane bonds and terminated with hydrophobic groups (see Fig 7). For a skilled research chemist, it is easy to guess how to synthesise such a structure. Actually, there are many synthetic routes claimed in the patent literature [32,33,34] and there is no need to present them all here. The simplest way is to react polyethylene glycol with diisocyanate to get a urethane prepolymer terminated with OH groups and then transfer it into a polyurethane of the structure presented in Fig 7, using monoisocyanate with a long hydrophobic group (eg dodecyl isocyanate).

Lengths of both hydrohpilic and hydrophobic segments, as well as their ratio, affect final properties of the product - solubility in water and thickening ability. It is believed [28] that the phenomenon of rapid increase in viscosity of water itself or water borno systems after addition of WSPU is a result of their specific molecular structure in water. Since they contain both hydrophilic and hydrophobic segments, they behave like surfactants in water. Hydrophobic segments form micelles and hydrophilic ones remain outside. As there are two hydrophobic segments in each molecule, the micelles are in fact linked with hydrophillic segments and the presence of such a crosslinked net of micelles in water results in increase in a viscosity. There are many theoretical and practical problems connected with this phenomenon eg the effect of various additives (electrolytes, solvents, low-molecular weight surfactants etc) on the thickening properties of WSPU in aqueous solutions and dispersions, but it is not the purpose of this paper to discuss them. What is, however, worth discussing here, is the perspective of using WSPU in water-borne adhesives.

In Fig 8, the viscosity of an acrylic dispersion (commercial product of Polish industry used as a binder for a dispersion lacquer for wood) before and after modification with two different WSPU synthesised in Author's laboratory is plotted vs shear rate. It is noteworthy that not only an increase in viscosity of this dispersion with the addition of WSPU, but also a distinct effect of shear rate on viscosity canbe observed. Since the only difference between both WSPU's was the ratio of hydrophillic/hydrophobic segment lengths, it seems that this is the factor responsible for thixotropic properties of water-borne systems thickened by WSPU. The practical conclusion from that for the adhesive formulator is to consider not only viscosity, but also thixotropic properties when using WSPU thickeners in adhesives.

Though in some papers and patents the use of WSPU in adhesives (especially acrylic) is described and their advantages over conventional thickeners (especially much higher effectiveness, better spreadability of thickened products and water-resistance of the films) are presented, we eventually chose commercial acrylic thickeners for our adhesives based on the blends of DPU and EVA dispersions we had developed for laminating PVC films to wood. This was because we noticed that the adhesion of the products was lower with WSPU compared to the other thickeners, eg acrylic [25]. The reason could be the presence of long hydrophobic

95

$$\text{HFB-U-}(\sim\!\!\sim\!\!\sim\!\!\sim\!\!\sim\!\!\sim\!\!\text{U-})_n\text{-HFB}$$

Fig 7   Structure of typical WSPU

∿∿∿ - hydrophilic chain
HFB - hydrophobic group
U - urethane bond

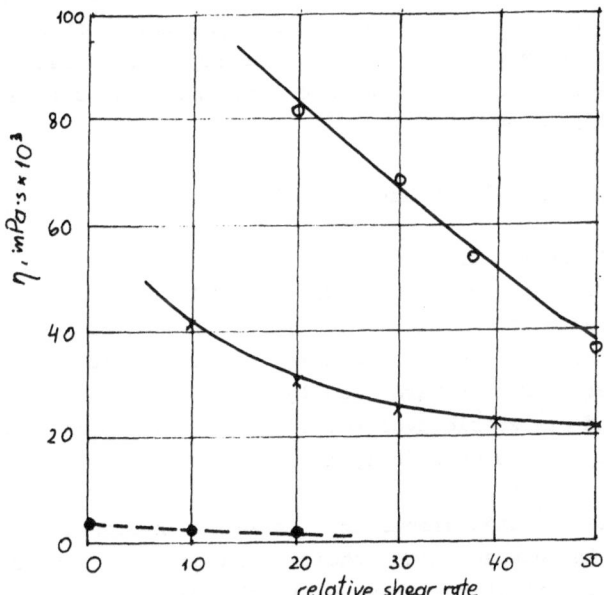

Fig 8   Viscosity of acrylic dispersion thickened with two
different WSPU vs shear rate

O - WSPU synthesised usinghigher ratio of
hydrophilic to hydrophobic segments

X - WSPU synthesized using lower ratio of
hydrophilic to hydrophobic segments

● - no thickener

groups in WSPU. Thereore, despite numerous positive features of WSPU thickeners, one must be rather sceptical about their potential use in water-borne adhesives. WSPU's are however, still developed and new formulations may suit better to adhesive applications.

Quite another future perspective, the possibility of using WSPU in adhesives can be, however, considered. This is a group of higher molcular weight WSPU's, without long hydrophobic units, which could be obtained, eg from polyethylene glycol or ethylene oxide-propylene oxide copolymers and diisocyanates, using a molar excess of OH. Such hydroxyl-terminated WSPU would presumably form viscous aqueous solutions and would perhaps be useful as moisture-activated adhesives. So far, there is no literature data on such products.

Water Dispersible Isocyanates (WDI)

It is well known that the isocyanates react with water forming ureas and carbon dioxide. The NCO-water reaction is, however, not too rapid, so the idea of preparing water-isocyanate dispersions does not look bad. Actually, such dispersions that are easily miscible with water-borne systems and are stable over several hours have been developed and even applied successfully in adhesives.

To make isocyanates dispersible in water, hydrophilic groups must be introduced into their structure. This can be realized by reacting them with the monoalkyl ether of polyethylene glycol [35] or with dihydroxy acids [7] (in the latter reaction partially blocked polyisocyanates are usually used). The general structure of the compounds obtained in these reactions is presented in Fig 9 a and b.

Compounds presented in Fig 9a are much more popular. In practice they are not used directly as water dispersibles isocyanates (WDI) but are added to a polyisocyanate that

becomes then water dispersible [36]. Of course, certain
modifications of this technology are possible. If an
isocyanate with partially blocked NCO groups is used in the
reaction with monoalkyl ether of polyethylene glycol [37] d
water isocyanate dispersion would be quite stable over very
long period of time because free NCO groups are blocked, but
the film made of it would cure easily at elevated air
temperatures in the presence of NCO-reactive components.

Aqueous dispersions of WDI containing free NCO groups mostly
based on MDI (4,4 - diphenylmethandiisocyanate), have already
been used for a couple of years as adhesive binders in
particleboard manufacturing instead of (or together with)
conventional urea-formaldehyde, melamine-formaldehyde or
phenol-formeldyhyde resins [38,39,40,41,42]. Here, NCO
reacts with both water and active hydrogen - containg groups
of wood. There are several remarkable advantages of such WDI
over conventional resin binders, the most important is that
the particleboards made using WDI do not contain formaldehyde
which is considered to be carciongenic and so they can be
used (eg in the furniture industry) without any health
restrictions. The other advantages are lower press curing
temperature, shorter press curing time, much better water
resistance. We proved this in our own experiments with 35%
aqueous dispersion of Suprasec 1042 (WDI based on MDI
supplied by ICI, UK)[43]. Someof the results are presented
in Table 2. Standard 57% urea formaldehyde resin with NH_4Cl
as curing agent was used for comparison.

TABLE 2

Results of testing WDI (Surprasec 1042, ICI) as binder for
particle boards. Results obtained for standard
urea-formaldehyde resin with NH CL as curing agent are
presented for comparison (Acc to [43])

Tested Properties	WDI	Urea-Form
Active substance content in the adhesive %	35	57
Adhesive content in the board %	6	10
Press-curing temperature °C	170	170
Press curing time minutes	3	6
Flexural Strength MPa	15.5	15.5
Tensile Strength MPa	0.78	0.54
Swell after 2 hours soaking in water %	9.2	11.6
Tensile Strength after 2 hours boiling in water (V-100 test) MPa	0.34	0

Another adhesives application of WDI is their use as
crosslinking agents for dispersion adhesives, not only for
those based on DPU described already in this paper, but also
for standard commercial acrylic and vinyl products as well as
for styrene-butadiene or polychloroprene lattices. They can
be also utilised as primers for fabrics, rubber, plastics,
concrete wood etc [44]. WDI containing carboxylic
functionality and made of partially blocked polyisocyanates
can be used as crosslinking agents for water-borne epoxy

adhesives. In this case, both carboxylic and unblocked isocyanate groups would react with epoxy resin at elevated temperatures [7]. It has been also reported that when the synthesis of NCO-terminated urethane prepolymer is carried out using polyester or polyetherdiol and WDI instead of conventional diisocyanate, the product can also be water dispersable [44]. Such aqueous dispersions of NCO-terminated prepolymers can be used as adhesives for bonding polyurethane foam, expanded vinyl elastomers, leather, wood etc. Bonding the foamed adherends seems to be especially interesting since the bond can be also foamed and the good soft handle of original material would not be spoiled.

As can be noticed from this brief review of polyurethanes and isocyanates with hydrophilic moieties, their possible adhesive application areas are still growing and undoubtedly some novel ideas in this field will appear in the near future. What still needs developing is a theoretical background for these systems, including both a chemical approach to their synthesis, crosslinking reactions etc, and a physical approach to their structure in water, adhesion behaviour etc.

a) $R-O-(-CH_2-CH_2-O-)_n-CO-NH-X-NCO$

b) $(BL-CO-NH)_2-X-NH-CO-\underset{COO^-}{\underline{}}-OC-NH-X-(NH-CO-BL)_2$

Fig 9   Structures of WDI

   (a) nonionic WDI containing isocyanate groups obtained from diisocyanate and monoalkyl ether of PEG
   (b) anionic WDI obtained from partially blocked triisocyanate and dihydroxy acid

      X - isocyanate moiety
      R - alkyl group (C - C)
        - dihydroxy acid moiety
      BL - NCO blocking

REFERENCES

1. Denton M.F., Chapter 11 in K.W. Allen "Adhesion 9", Elsevier Appl Sci Publ, London New York 1984

2. Goldsmith J., Coated Fabrics 18 (July 1988) p12

3. Dieterich D., Progr Org Coat 9, 281(1981)

4. Dieterich D., Proc of Third Int Conf in Organic Coatings and Technology vol 1

5. Deiterich D., Angew Makromol Chem 98, 133 (1981)

6. Tirpak R.E., Markusch P.H., J Coat Technol 58(738) 49 (1986)

7. Rosthauser J.W., Nachtkamp K., J Coated Fabrics 16 (July 1986) p39

8. Rosthauser J.W., Williams J.L., Proc of ACS Div of Polym Mat Sci Eng 50, 344 (1984)

9. Schmelzer H.G., J Coated Fabrics 17 (January 1988) p167

10. Tsirovasiles J.C., Tyskwicz A.S., ibid 16 (October 1986), p114

11. Prelini C., Trovati A., Turchini L., Proc of UTECH'88 Conference, The Hague OCtober 1988 p220

12. Reich F., Textilveredlung 13, 454 (1978)

13. Traubel H., Leder 28, 181 (1977)

14. Lorenz D., Reinmoller K.H., Kautsch Gummi Kunst 34, 827 (1981)

15. Lorenz D., Reinmoller K.H., ibid 35, 825, (1982)

16. Japan Pat J 54 127 442

17. Germany Pat 1 953 348

18. Kujawa Penczek B., Mielniczuk E., Kozakiewicz J., unpublished results

19. Japan Pat 61 287 972

20. Spain Pat 513 375

21. Brit Pat 1 191 260

22. German Pat 2 804 609

23. Japan Pat 80 08 344

24. Britain Pat 1 250 266

25. Kozakiewicz J., Koncka-Foland A., unpublished results

26. Maempel L., Adhesion No 5 (1988) p14

27. Daniel P., Quinn K.J., Proc of the Int Conf on Footwear Materials Tschechoslokia (1984)

28. Bieleman J.H., Riesthuis F.J.J. Van der Velden P.M. Farbe + Lack 93, 570 (1987)

29. Glass J. E., Fernando R.H., England-Jongewaard S.K., Brown R.G., JOCCA 67(10) 256 (1984)

30. Glass J.E. Am Paint Coat J., August 1984 p45

31. Glancy C.W., ibid p44

32. US Patent 4 426 485

33. US Patent 4 155 892

34. US Patent 4 496 708

35. Britain Patent 1 417 618

36. Britain Patent 1 144 933

37. US Patent 4 098 933

38. Ball G.W., Brown J.P., McDermott P., Hennessey P.M. Paper presented at FESYP Conference October 1982

39. Gaul J.M., Nguyen T., Babiec Jr J.S., J Elastomers and Plastics 16, 206 (1984)

40. Pagel H.F., Luckman E.R., JAPS Appl Polym Symp 40, 191, (1984)

41. US Patent 4 431 455

42. US Patent 4 306 673

43. Pawlicki J., Onisko W., Dobrowolska W., Lendzion A., Kozakiewicz J., 5th Scientific Conf on Wood Products Agricultural Univ Warsaw, November 1989

44. Robertson J.R., Cohen S.C., McNellis K.M. ibid 15, 113 (1983)

7

INTERACTION OF CHROMIUM COMPOUNDS WITH POLYURETHANES

J Gahde, I Loeschcke, J Sachse, W Hiller,
G Schulz, H Goehring

Academy of Sciences of the GDR
Central Institute of Organic Chemistry
Rudower Chausse 5
DDR-1199 Berlin

1. INTRODUCTION

The properties of a filled polymer depend on the properties of the polymer, the properties of the filler and the interaction between the two components [1,2]. Many of the polymeric coating materials are filled with inorganic pigments. Special types of coating systems, for example magnetic recording materials, are highly filled polymers [3]. The addition of any mineral filler increases the modulus and the temperature deflection of the polymer, but also the brittleness.

The nature of the interface between filler and polymer plays an important role in determining many aspects of the composite efficiency. In the first stages of coating preparation the pigment has to be dispersed in the polymeric solution by adding a dispersant, which is a surface active substance and modifies the surface of the pigment in the suspension and later on in the coating. The second reason for the application of such a surface modifier is to maintain the polymer-pigment interaction and to preserve the mechanical properties of composites in adverse environmental conditions. To reduce the brittleness of a highly pigmented polymer, however, it is necessary to produce a highly elastic polymeric matrix [4,5,6]. If the elastic modulus E_x of the polymeric interlayer is much lower than the elastic modulus E_m of the polymeric matrix, the modulus of the composite decreases with increasing filler content. For such a behaviour it is necessary that the thickness of the interlayer exceeds 100nm. Furthermore, an adhesion interaction between the polymeric interlayer and the components of the composite is required. However, such a structure of the composite can only be realised with a chemical interaction between the pigment and the polymeric interlayer, which changes the structure and properties of the polymer in the interlayer. In a CrO_2 filled polyurethane/VC polymer coating system, where the polyurethane is mainly attached to CrO_2, we have such a situation. The properties of the highly elastic polurethane are changed by chemical interaction with CrO_2 and an explanation for that behaviour will be given.

2. EXPERIMENTAL

Materials

The investigations were carried out on two different polyurethanes composed as follows:

Type I   PUR Mn ca 50000 polyester
             polytetramethylane adipatediol Mn ca 1000
             4,4 di-phenylmethane diisocyanate (MDI)
             1,4 butanediol (chain extender)

Type II   PUR Mn ca 30000
              polyester (polymethylene adipatediol) Mn ca 800
              MDI
              neopentylglycol (chain extender)

The main difference between the two polyurethanes was in the use of a sterically hindered alcohol as chain extender in Type II. The chromium dioxide had been produced by VEB Filmfabrik Wolfen, GDR, with a specific surface of 25 m^2 /g.

Formation of Coatings

The polyurethanes were dissolved in methyl ethyl ketone, to give a solution containing 5 wt. % polyurethane. The solution was poured into a roll mill, CrO_2 added and dispersed for 48 hours at room temperature. Afterwards the suspension was poured onto a glass plate to create 100um thick coating. The drying of the film was carried out at first at room temperature, and then for 3 hours in the temperature range 30-40° C under vacuum (ca 10Pa). The dried film was removed from the glass and stored as follows: 7 days, 70° C, relative humidity 95%. The concentration of CrO_2 in the film was 67 wt-%.

Methods

For characterisation of the molecular mass, the films were dissolved in tetrahydrofurance before and after storage and the clear solution was investigated by size exclusion chromotography (SEC) related to polystyrene as a standard. The columns were filled with PS20 and PS4000, produced by Merck, FRG.

The XPS or ESCA measurements were carried out with a spectrometer by Kratos, type ES200B. The radiation was Al K (hv = 1486.6eV). The CrO_2 powder was dispersed on a tape of the sample carrier of pressed into a tablet. The basic peak was the C1s peak of the CH_2 and CH groups (BE 285.0 Ev).

For the NMR measurements we used a Bruker-MSL 400 NMR Spectrometer. The concentration of polymer solution was 4 wt -% for ^1H measurements and 30 wt-% for ^{13}C investigations. Five mm tubes were used in the spinning mode with acetone-d as solvent and hexamethyldisiloxane (HMDS) as internal reference.

To determine the Young's modulus E we recorded the E/T diagrams with a Viscoelastometer Rhevibron, Model DRV-II-B of Toyo Baldwin Co Japan. The films were 0.3 x 5 x 30mm. The thickness of samples was produced by laying 5 films one upon each other. Measurements were carried out at a frequency of 3.5Hz.

Results and Discussion

Chromium dioxide is a magnetic pigment which decomposes into chromium III and chromium VI compounds in the presence of water [7]. This disproportionation occurs in the following way:

$$3 \ CrO_2 + 4 \ H_2O \underset{>100^\circ C}{\overset{RT}{\rightleftharpoons}} H_2CrO_4 + 2 \ Cr(OH)_3 \downarrow$$

$$Cr_2O_3 + 3H_2O$$

Other chromium oxides, such as CrOOH or chromium with a valency between three and six may be formed too [8]. The results obtained from XPS investigations agree with the reaction mechanism discussed. The main products at the surface of chromium dioxide are chromium III and chromium VI compounds in the relation mentioned above two to one. Nearly 20% of the chromium compounds found cannot be attached directly to a distinct chromium oxide compound. Therefore,

the interactions of chromium dioxide should be of physical
and chemical nature as shown in Scheme 1.

It is known that chromium III compounds are destined for
complex formation with amines or nitrogen containing
compounds [9]. Consequently our work is aimed at
investigating interactions of chromium III chloride and
polyurethanes, where the chromium III chloride is partly
soluble in acetone and, therefore, a suitable model for
studying the interaction of a chromium III compound with
polyurethanes. The ^2H spectra of polyurethanes have shown a
broadening of all signals with increasing chromium III
content. A significant decrease of NH intensities of both
polyurethanes has been observed compared with aromatic
signals. This fact is confirmed in Table 1, where the ratios
of the integrals of NH/aromatic signals are presented.

The ^{13}C nmr measurements confirm the observation that there is
a strong interaction between the NH of the urethane group and
chromium III, as shown in Fig 1. The chemical shift of the
CH   groups in the neighbourhood of the ester group shows a
great chemical shift as well. Therefore, NMR measurements
have proved an interaction of chromium III chloride with NH
and CO groups in polyurethanes. This is caused by an
intramolecular and intermoleculr interaction and should
result in a crosslinking of polyurethanes on addition of
chromium III chloride. The proof is given by DMA
measurements (Fig 2). By incorporation of 20 wt-% chromium
III chloride in the polyurethane II the elastic modulus forms
a plateau in the temperature range from 100 to 200°C. The
chromium III chloride acts as a chemical crosslinker as shown
in Fig 2. By complex formation chromium III compounds
strengthen the polyurethanes and gives the film a higher
modulus and thermal stability.

The second point involves a chromium VI compound at the
surface of CrO_2 . From the disproportionation equation the
compound should be chromic acid. After the suspension of

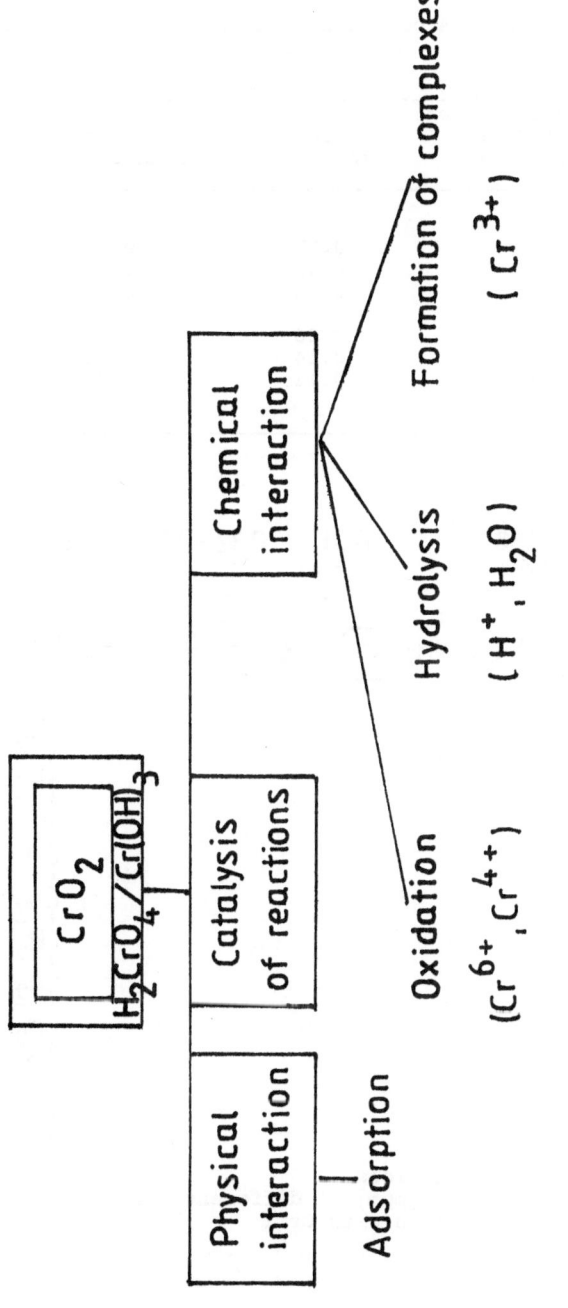

Scheme 1 Interaction of chromium dioxide

TABLE 1
Integral ratio of NH/Aromatic signals of 1H nmr measurements
n = INH/Iarom

Sample	Concentration of CrCl3 (wt.-%)	n fresh solutions	n solutions 3 days stored
PUR I	-	0.22	0.7
PUR I	13	0.15	0.04
PUR I	38	0.16	0.07
PUR II	-	0.20	0.19
PUR II	20	0.14	0.12
PUR II	50	0.12	0.06

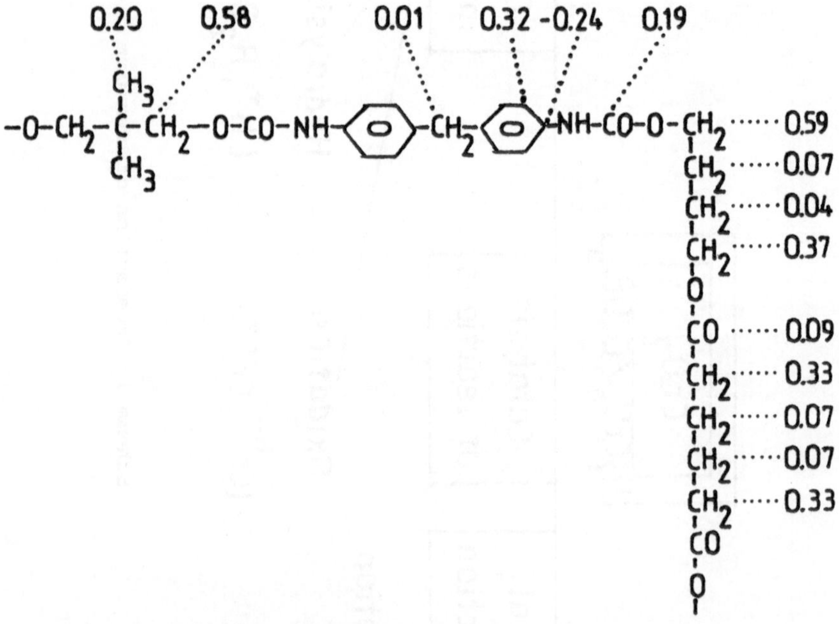

Fig 1: C NMR investigations
Chemical shift (ppm) of different carbon atoms in PUR
II by interaction with CrCl

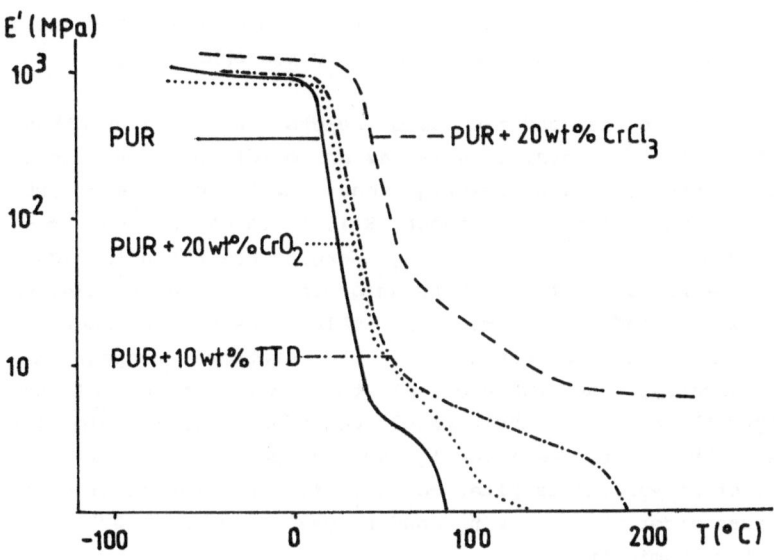

Fig 2: Dynamic mechanical behaviour of PUR II by addition of the chemical cross linker TTD or chromium III chloride

$$TTD \quad CH_3 - CH_2 - C \left\langle \begin{array}{l} CH_2 - O - OC - HN - \text{(O)} - NCO \\ CH_2 - O - OC - HN - \text{(O)} - NCO \\ CH_2 - O - OC - HN - \text{(O)} - NCO \end{array} \right.$$

chromium dioxide in distilled water, a pH from 2.6 to 3.3 has been found showing a strong acid reaction. That points to the formation of chromic acid at the surface of CrO2. Furthermore, the concentration of chromic acid has been measured by oxidation of benzhydrol to benzophenone [10].

It is assumed that chromic acid reacts in a polyurethane system into two ways, firstly as an oxidising agent as we could observe by XPS investigations. At the surface of CrO2 a polyurethane layer of about 8-10nm thickness has been adsorbed. The observed bonding energies are shown in Figure 4. The nitrogen of the urethane group is partially oxidised to nitroso and nitro compounds. This is assumed to occur in connection withthe rupture of the urethane linkage and consequently, the rupture of the main chain of the polyurethane. The result should be a decrease in molecular mass. Therefore, we investigated changes in the molecular mass of polyurethanes after contact with chromium dioxide and storage under the following conditions: 7 days, 70 $^{\circ}$, 95% relative humidity.

The SEC investigations show a high decrease in molecular mass after storing the samples under the above mentioned conditions (Fig 3 and 5). PUR II seems to be more stable than PUR I. The reason can be sterically hindered alcohol (neopentylglycol) as chain extender. If the initial molecular mass M_n of the two polyurethanes is equal to 100% after storage M_n of PUR I is about 32% and PUR II about 61%.

In PUR I the reduction of molecular mass occurs during the preparation of the suspension and the film formation, ie before storage under climatic conditions. Figure 4 shows that a surface modification of chromium dioxide by treatment with Lecithin does not hinder the decrease in molecular mass. It is supposed that the ester linkage will break by saponification too, but at present time we do not have results about this. Surprisingly, the increase in molecular mass was mainly at PUR I after contact and storage with

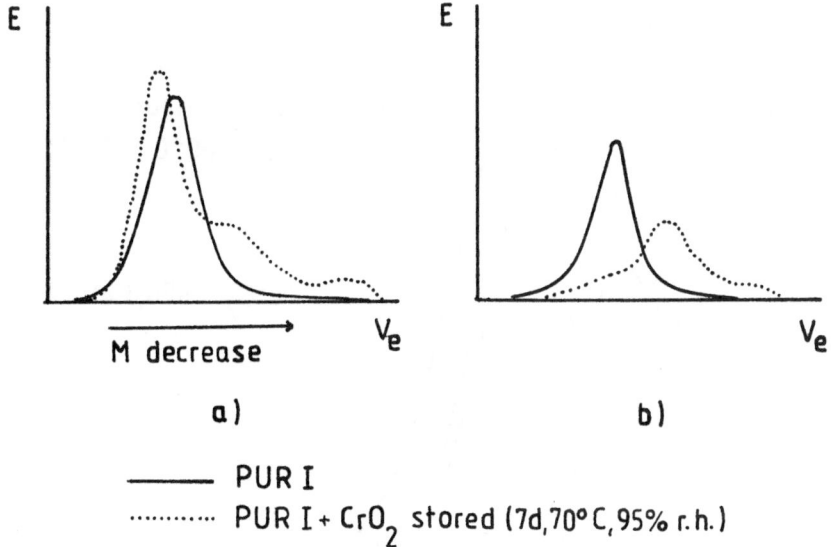

Fig 3: SEC Investigations of PUR I (E=UV extinction, V=
        evolution volume)
        Changes in molecular mass of PUR I:
        (a) uncoated
        (b) coated with 5 wt-% lecithin
        after interaction with chromium dioxide

Sample	Mw	Mn	U
PUR I	121000	45700	2.6
PUR I + CrO	124000	22600	5.5
stored 7 days, 70 , 95% RH			

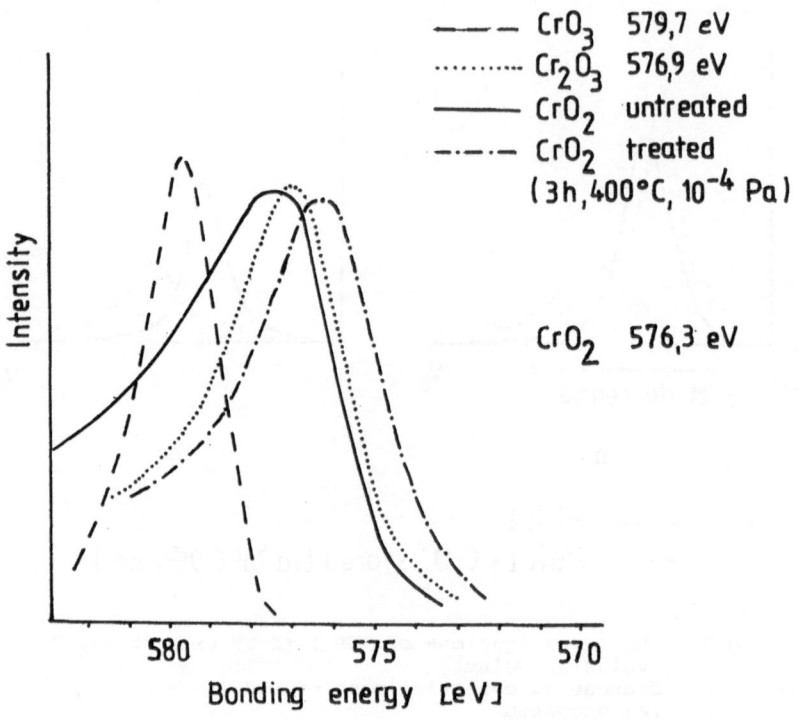

Fig 4: SEC Investigations of (a) PUR I and (b) PUR II

a) Sample	Mw	Mn	U
PUR I	114000	62000	1.8
PUR I + CrO	86000	18000	4.7
PUR I + CrO stored 7 days, 70 , 95% RH	120000	20000	5.5

b) Sample	Mw	Mn	U
PUR II	53300	30000	1.8
PUR II + CrO	61000	19000	2.1
PUR II + CrO stored 7 days, 70 , 95% RH	86400	18300	4.7

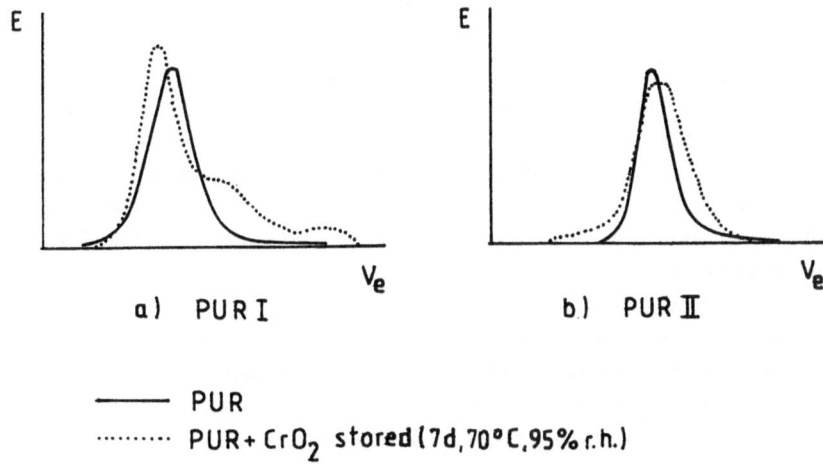

a) PUR I   b) PUR II

——— PUR
········· PUR+ CrO_2 stored (7d, 70°C, 95% r.h.)

Fig 5: SEC investigations of PUR I
Influence of solvents on changes of molecular mass.
(a) tetrahydrofurane, (b) dichloroethane

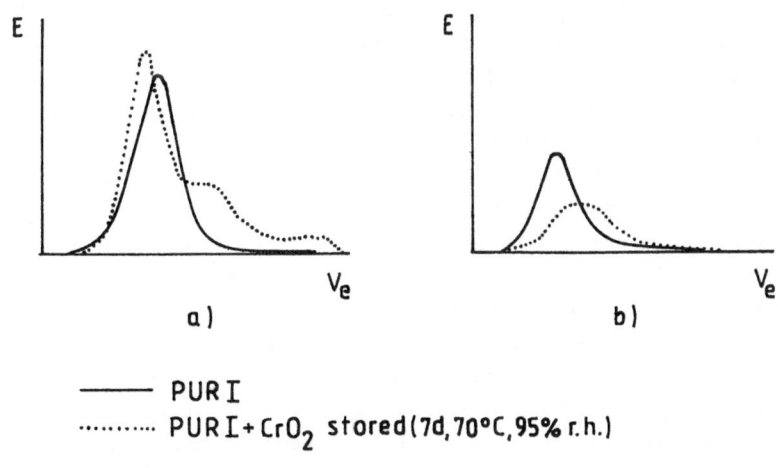

a)   b)

——— PUR I
········· PUR I+ CrO_2 stored (7d, 70°C, 95% r.h.)

Fig 6: SEC Investigations of PUR I
Influence of solvents on changes of molecular mass.
(a) tetrahydrofurane
(b) dichloroethane

chromium dioxide. The reason for that can be a
polymerization initiated by peroxide radicals during storage
at 70^{0}C. We found an increase in the peroxide content up to
0.3% in the solvents investigated, tetrahydrourane,
cyclohexanone and methyl ethyl ketone and it was possible to
polymerize styrene at room or elevated temperatures. The
polymerization occurs to a greater extent in the presence of
chromium dioxide. That means chromium dioxide catalyzes the
perioxide and/or the radical formation. In the solvent 1.2
dichloroethane we could not observe the formation of
peroxides during treatment with chromium dioxide. Therefore,
the increase in molecular mass is depressed by using this
solvent for film preparation as shown in Fig 6. Further
investigations will elucidate this problem in more detail.

Summary and Conclusions

Chromium dioxide is a chemically active pigment. It
disproportionates mainly into chromium III and chromium VI
compounds as shown by XPS investigations. Chromium III
chloride as a model for a chromium III compound forms
complexes with polyurethanes by interaction with NH and CO
groups. The formation of chromic acid at the surface of
chromium dioxide is the main reason for acidic and oxidising
reactions of the pigment. The structure of polyurethanes is
considerably changed by interaction with chromium dioxide. A
decrease in molecular mass is observed after storing
CrO_2 filled polyurethanes caused by oxidation of the
nitrogen inthe urethane group and presumably by
saponification of the ester group too. The observation of an
increase in molecular mass was surprising. This phenomenon
is under investigation now. First results show a connection
wih peroxide formation in the solvents used and the
possibility of radicalic reactions and polymerization. PUR
II contains a sterically hindered alcohol as a chain extender
and in contact with chromium dioxide it is more stable than
PUR I. We suppose that sterically hindered alcohol both in

the polyester of the soft segment and as a chain extender in the hard segment will afford an even more stable polyurethane.

Acknowledgements

The authors would like to thank the following people for their assistance in measurements and discussions: R. P. Kruger (chemical analysis), R. Gehrke (sample preparation), H. Becker (GC)

REFERENCES

1. Sheldon, R.P., "Composite Polymer Materials" Applied Science Publishers, London New York, 1982.

2. Berlin, A. A., St A Volfson, Enikolopian N.S., Negmatov, Principles of Polymer Composites, Akademie-Verlag Berlin, 1986.

3. Braginskikif, G.I., Timofeev, E.N., Technologie der Magnetbandherstellung, Akademie-Verlag Berlin, 1981.

4. Luders, G., Carius H.E., Mechanische Eigenschaften von gefulltem kautschukmodifiziertem Polystyren Plaste u Kautschuk 26,10 (1979) p563.

5. Gahde J., Zur Bedeutung der Phasengrenz bzw Zwischenschicht fur mecahnische Eigenschaften anorganisch gefullter Polymere.
   Promotion B Akademie d. Wissenschaften der DDR.

6. Gahde, J., About Interface Problems in Kaolin-Filled Polythylene Polymer Composites Editor Sedlacek B. Walter de Gruyter Berlin-New York 1986, p431.

7. Garbassi F., MelloCeresa E., The Mechanism of the Surface Stabilisation of CrO2 Magnetic Powers Applications of Surface Science 14 (1982-83) p330.

8. Khilla M. A., Hanafi Z.M., Mohamed A. K., Infrared Absorbtion of Chromiumtrioxide and its Suboxides Thermo chimica Acta 54 (1982) p319.

9. Cotton, F. A., Wilkinson G., Anorganische Chemie VEB Deutscher Verlag der Wissenschaften Berlin 1968, p770.

10. US Patent 3512530 C 01 G 37/02
    Du Pont 7.5.1969

8

SESSILE DROPS ON HETEROGENEOUS SURFACES :
STATIC AND DYNAMIC BEHAVIOUR

M.E.R. SHANAHAN
Centre National de la Recherche Scientifique
Ecole Nationale Supérieure des Mines de Paris
Centre des Matériaux P.M. Fourt
B.P. 87
91003 EVRY CEDEX - FRANCE

INTRODUCTION

A sessile drop of a pure liquid, L, resting on a perfectly flat, smooth horizontal, isotropic, homogeneous and undeformable solid, S, in the presence of a second fluid, which we shall take to be the vapour V of L for definiteness, should adopt an axisymmetric shape at equilibrium. However, it is an everyday observation that the vast majority of drops on solid surfaces take a form which is anything but axisymmetric. This is related to the phenomenon of wetting hysteresis, or potential variability of the contact angle, θ, (subtended in the liquid phase between the solid surface and the tangent to the liquid/vapour interface) generally caused by inhomogeneity of the surface. A degree of solid heterogeneity may arise from one or several of a host of possible causes including the presence of chemical inhomogeneities, local specific adsorption, surface roughness, molecular orientation and the existence of local solid deformation in the vicinity of the triple line [1-7]. Whatever be the essential reason(s) for this contact angle hysteresis, chemical or physical, the fact remains that the simple meniscus shape corresponding to the circular contact line of an axisymmetric drop becomes deformed and solution of Laplace's capillary equation [8] is then, in general, rather difficult. This relation, describing the equilibrium of the liquid/vapour interface, is a second-order, non-linear partial differential equation which rarely affords an analytical solution (not even in the axisymmetric case if gravitational effects are to be taken into consideration [9]).

Despite the complexity of the general problem described above, the aim of the present study is to consider the shape of liquid menisci

116

perturbed by solid surface inhomogeneities. In order the render the mathematics tractable, we shall make a few simplifying assumptions in the calculation, notably that gravitational effects are negligible (small drops), that the intrinsic, unperturbed value of the contact angle, θ_o, is small, and that the field of heterogeneities, although random, is relatively weak.

A consequence of the analysis is it can be shown that in most cases, such a perturbed drop is not strictly at equilibrium and slow movement of the drop as a whole occurs in the search for a thermodynamically stable configuration. A tentative comparison of this process with cellular phagocytosis is made.

THEORY OF DROP SHAPE

A more complete description of the theory is given elsewhere [10] and so the following will be limited to basic principles and equations.

Free energy values γ_{sv}, γ_{sL} and γ are attributed respectively to the three interfaces solid/vapour, solid/liquid and liquid/vapour. An initial drop resting on the solid and of the form of a spherical segment (no gravity or heterogeneity effects) with radius of curvature R and contact radius r_o ($\sim R\,\theta_o$) has at its centre the origin of polar coordinates (r, Φ) in the plane of the solid surface (see Fig.1). This drop is (mathematically) perturbed by a random additional free energy of the solid surface, ϵ (r, Φ), which modifies the value of γ_{sv} to $(\gamma_{sv} + \epsilon)$. Taking drop height with respect to the solid surface as z (r, Φ), we may write an expression for the total free energy of the system, E, together with the constraint of constant liquid volume, V :

$$J = E - \Delta p. \, V \sim$$

$$\int_{-\pi}^{+\pi}\int_{0}^{p(\Phi)} \left\{ \gamma\left(1 + \frac{z_r^2}{2} + \frac{z_\phi^2}{2r^2}\right) + \gamma_{sL} - \gamma_{sv} - \epsilon - \Delta p.z \right\} r \, dr \, d\Phi + \text{constant} \quad (1).$$

In equation (1), Δp is the Laplace pressure across the liquid/vapour interface $(\Delta p = 2\gamma/R)$, $\rho(\Phi)$ is the modified contact radius $[\rho\,(\Phi) = r_o + \delta\,(\Phi)]$ and z_r, z_ϕ have their usual meanings as derivatives. In the following, ϵ is taken to be weak $(|\epsilon|/\gamma \ll 1)$ and $(\gamma_{sL} - \gamma_{sv})$ corresponds to an average value $(\int_A \epsilon \, dA \sim 0$ where A is the solid surface area considered).

For the system to be at equilibrium, J must be minimal and thus the functional defined by equation (1) must satisfy Ostrogradskii's equation

[11] of the variational calculus. In addition, we note that due to the low intrinsic contact angle, θ_o, and the fact that we shall "perturb" the spherical segment of the initial drop, $h(r)$, by a small function, $\eta\ (r, \phi)$, we may write :

$$z\ (r, \phi) = h(r) + \eta\ (r, \phi) \sim \frac{(r_o^2 - r^2)}{2R} + \eta\ (r, \phi)\ ;$$

$$|\eta/r_o|,\ |\eta_r|,\ |\eta_\phi/r_o|\ \ll 1 \tag{2}.$$

Incorporating equation (2) into equation (1) and using Ostrogradskii's equation, we obtain after simplification Laplace's differential equation for η :

$$r^2\ \eta_{rr} + r\eta_r + \eta_{\phi\phi} = 0 \tag{3}.$$

This may be solved by standard methods [12] leading to :

$$\eta\ (r, \phi) = \frac{\beta_o}{2} + \sum_{n=1}^{\infty} r^n\ (\alpha_n \sin n\phi + \beta_n \cos n\ \phi) \tag{4},$$

where α and β represent constants to be obtained from the boundary conditions of the problem.

Having obtained the functional form of η using variational techniques, we may determine α and β using standard techniques of the differential calculus. Since equation (4) is essentially a Fourier series, we express the perturbation field in a similar form, realising that the only contributions of importance are those near the triple line :

$$\epsilon\ (r_o, \phi) = \frac{B_o}{2} + \sum_{m=1}^{\infty} (A_m \sin m\ \phi + B_m \cos m\ \phi) \tag{5}.$$

Use of equations (2), (4) and (5) allows us to re-express equation (1) in the following form :

$$J \sim \pi \left\{ \frac{\gamma}{4} \left[r_o^2\ \theta_o^2 - \beta_o^2 + 2 \sum_{n=2}^{\infty} (n-1)\ r_o^{2n} (\alpha_n^2 + \beta_n^2) \right] \right.$$

$$\left. - R \left[\frac{\beta_o B_o}{2} + \sum_{n=1}^{\infty} r_o^n\ (\alpha_n A_n + \beta_n B_n) \right] \right\} \tag{6}.$$

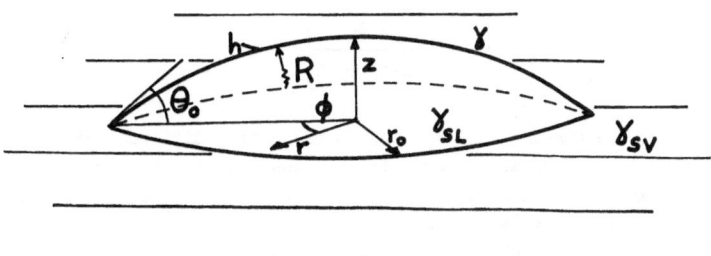

Figure 1. Sessile drop of small intrinsic contact angle, θ_o.

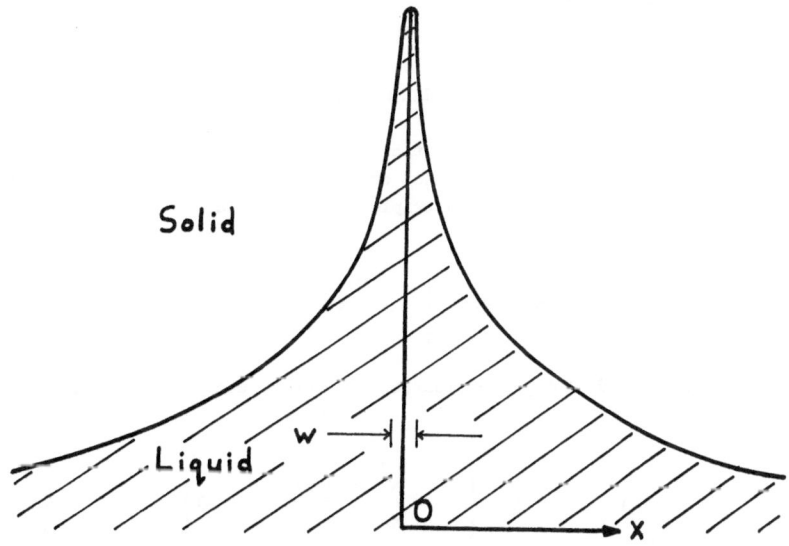

Figure 2. Representation of a logarithmic "spur" on the triple line produced by a local heterogeneity.

The differential of J with respect to each α and β must be zero for equilibrium to ensue. This leads to the following values :

$$\left.\begin{array}{ll} \alpha_n & = \dfrac{R\ A_n}{(n-1)\ r_o^n\ \gamma} \\[4mm] \beta_n & = \dfrac{R\ B_n}{(n-1)\ r_o^n\ \gamma} \end{array}\right\} \tag{7},$$

except for the cases α_1 and β_1. The corresponding differentials are zero only when A_1 and B_1 are zero. It is thus only in these circumstances that we may assume the drop to be strictly at equilibrium. Use of the expressions (7) and equations (2) and (4) give the meniscus shape. Similarly, modifications, δ, to the unperturbed contact radius, and to the contact angle, $\Delta\theta$, may be assessed :

$$\delta\ (\phi) \sim \frac{r_o}{\gamma\ \theta_o^2}\ \left\{ -\frac{B_o}{2} + \sum_{n=2}^{\infty} \frac{1}{(n-1)}\ [A_n\ \sin n\phi + B_n\ \cos n\phi] \right\} \tag{8},$$

$$\Delta\theta\ (\phi) \sim \frac{1}{\gamma\theta_o}\ [\epsilon\ (r_o,\ \phi) - A_1\ \sin\ \phi - B_1\ \cos\ \phi] \tag{9}.$$

A SINGLE PERIPHERAL HETEROGENEITY

Having obtained the general formulation, we shall consider a particularly simple case of its application, that corresponding to a single, isolated (positive) heterogeneity near the triple line such that :

$$\epsilon\ (r_o,\ \phi) = \begin{cases} 0 & ; \quad -\pi < \phi < -\chi \\ \epsilon_o & ; \quad -\chi < \phi < \chi \\ 0 & ; \quad \chi < \phi < \pi \end{cases} \tag{10}.$$

Standard methods for the evaluation of Fourier coefficients lead to :

$$\epsilon\ (r_o,\ \phi) = \frac{\epsilon_o}{\pi}\ \left\{ \chi + 2 \sum_{n=1}^{\infty} \frac{\sin n\chi}{n}\ .\ \cos n\phi \right\} \tag{11},$$

whence we obtain :

$$\delta(\phi) \sim \frac{r_o\,\epsilon_o}{\pi\,\gamma\,\theta_o^2}\left[-\chi + 2\sum_{n=2}^{\infty}\frac{\sin n\,\chi}{n(n-1)}\cdot\cos n\,\phi\right]\qquad(12).$$

Although, we have an expression for the protuberance, or "spur", on the drop periphery caused by ϵ, its series representation makes it a little awkward to handle. However, we may simplify this considerably. Since both ϵ_o and χ will generally be relatively small compared to r_o, we let $r_o \longrightarrow \infty$(mathematically). In this case, equation (12) can be written :

$$\delta(\phi) \sim \frac{r_o\epsilon_o\,\chi}{\pi\,\gamma\,\theta_o^2}\left[-1 + 2\sum_{n=2}^{\infty}\frac{\cos n\,\phi}{(n-1)}\right]\qquad(13).$$

Since the values of ϕ of interest are small, we have $\cos n\,\phi \sim \cos\phi\cdot\cos[(n-1)\phi]$. Using the trigonometric identity [13] :

$$\sum_{n=1}^{\infty}\frac{\cos n\,\phi}{n} = \ln\left(\frac{1}{2}\operatorname{cosec}\frac{\phi}{2}\right)\quad;\;0<\phi<2\,\pi\qquad(14),$$

it may than be shown that :

$$\delta(\phi) \sim \frac{r_o\,\epsilon_o\,\chi}{\pi\,\gamma\,\theta_o^2}[-1 - 2\ln|\phi|] \sim \frac{-2\,r_o\,\epsilon\,\chi}{\pi\,\gamma\,\theta_o^2}\ln|\phi|\qquad(15).$$

The effective force, f, exerted by ϵ radially away from the centre of the drop is equal to $2\,r_o\epsilon_o\,\chi$ and in the limit as $r_o \longrightarrow \infty$, the polar angle ϕ may be replaced by x/r_o where x is the distance measured along the (undisturbed) triple line from the centre of ϵ. Equation (15) may then be written alternatively as :

$$\delta(x) \sim \frac{f}{\pi\,\gamma\,\theta_o^2}\left[-\frac{1}{2} + \ln\frac{r_o}{|x|}\right] \sim \frac{f}{\pi\,\gamma\,\theta_o^2}\ln\frac{r_o}{|x|}\qquad(16),$$

where the range of validity is restricted to $|x| \gtrsim w/2 \sim r_o\,\chi$.

In order to find the actual length of the spur, i.e. $\delta(0)$, a similar argument is applied, first evaluating at $\phi = 0$, before effecting the summation. The process leads to :

$$\delta(0) \sim \frac{f}{\pi\,\gamma\,\theta_o^2}\left[\frac{1}{2} - \ln\chi\right] \sim \frac{f}{\pi\,\gamma\,\theta_o^2}\ln\frac{2\,r_o}{w}\qquad(17).$$

A representation of a logarithmic spur on the triple line is given in Fig.2.

Both equations (16) and (17) have been obtained by Joanny and de Gennes [14] using a different approach (in which the triple line is assumed straight at the outset). These same expressions show that the difference between δ (0) and δ (x) is relatively insignificant thus indicating that the width of the heterogeneity, w, serves effectively as a cut-off for the essentially logarithmic deformation of the triple line caused by the presence of ϵ.

Equations (12), (16) and (17) may be generalised for cases corresponding to the presence of multiple heterogeneities near the triple line.

It may be noted that for the case studied, corresponding to a single peripheral heterogeneity, the drop is in fact in a state of quasi-stable equilibrium (i.e. the meniscus shape corresponds to a state of minimal free energy for the specified, liquid/solid contact zone). This becomes clear either from the fact that B_1 is finite in the series representation of ϵ in equation (11) ($B_1 = 2 \epsilon_0 \pi^{-1} \sin x$) or alternatively by observing that the isolated force, f, produced by the heterogeneity, ϵ, is not equilibrated by effects from other inhomogeneous zones. The net force acting on the drop, although small, will lead to drop motion as shown in the next section.

AN UNEQUILIBRATED HETEROGENEITY : DROP MOTION

The existence of A_1 and/or B_1 terms in the Fourier series for the field of heterogeneities near the drop periphery (equation (5)) modifies the contact angle in such a way that Young's equation is not respected. Were perturbations to lead to a strictly equilibrated state, local modification to the contact angle would be given by :

$$\Delta\theta \ (\phi) \sim \frac{\epsilon \ (r_o, \phi)}{\gamma \ \theta_o} \qquad (18).$$

Comparison with equation (9) makes the point. Thus one way to estimate the overall force tending to move the drop is to integrate the modified Young equation (leading to an imbalance of capillary forces) round the triple line [15].

An alternative method, in the case of an isolated heterogeneity, is to use the expression (17) inverted to give the force, f, explicitly :

$$f \sim \pi \gamma \theta_o^2 \, 1^{-1} \, \delta \, (0) \qquad (19),$$

where $1 = \ln \, (\, 2 \, r_o/w)$. We shall now consider that the spur is of decreasing length since, as the drop is "pulled" by f, the heterogeneity, ϵ, becomes increasingly covered by the liquid during engulfment. Both f and δ (0) diminish and adopting $\delta(t)$ to represent henceforth the value of the length the spur (equation (17)) given as a function of time, t, we have :

$$f \, (t) \sim \pi \gamma \theta_o^2 \, 1^{-1} \, \delta \, (t) \qquad (20).$$

If the drop moves as a whole at speed U, the work done by f (per second) is Uf.

Viscosity Effects

As the drop moves under the influence of f, energy is dissipated within the liquid due to viscous drag. The velocity field within the drop will be complex, but we may vastly simplify the calculation by adopting the lubrication approximation applicable to cases of small contact angle [16]. The drop is taken to be a spherical segment (η of equation (2) is assumed negligible compared to h). Flow direction is considered to be uniquely parallel to the solid surface with a parabolic, vertical velocity gradient, $u(z)$. Boundary conditions of no slip at the liquid/solid interface and no tangential stresses at the liquid/vapour interface are used together with an overall flux corresponding to speed U. [7, 16]. Despite the relatively simple flow field, allowance must be made for a cut-off at distance i from the periphery within the drop, otherwide singularities appear (due to infinite velocity gradients). The velocity gradient obtained is :

$$u \, (z) \sim \frac{3U}{2 \, h^2} \, (2 \, h \, z - z^2) \qquad (21).$$

Using the cartesian coordinates of Fig.3, we see that dissipation will arise from velocity gradients ($\partial u/\partial z$) and ($\partial u/\partial x$) : the situation is more complicated than that of pure radial flow where only the former gradient exists (in the elementary analysis).

$$\frac{\partial u}{\partial z} \sim \frac{3\ U}{h^2}\ (h - z) \tag{22}$$

$$\frac{\partial u}{\partial x} \sim \frac{3\ U}{h^2}\left(\frac{z^2}{h} - z\right)\frac{\partial h}{\partial x} \tag{23}$$

$$h \sim \frac{1}{2R}\left(r_o^2 - x^2 - y^2\right) \tag{24}.$$

Employing equations (22) to (24) and making a few suitable simplifications, we may calculate the viscous dissipation (per second), $T\dot{S}$, in the drop volume, V :

$$T\dot{S} \sim \mu \int_V \left[\left(\frac{\partial u}{\partial z}\right)^2 + \left(\frac{\partial u}{\partial x}\right)^2\right]\ dV \sim 3\ \pi\ \mu\ U^2\ r_o\ L\left(\frac{2}{\theta_o} + \frac{\theta_o}{10}\right) \tag{25},$$

where μ is viscosity and $L = \ln\ (r_o/i)$. It is interesting to note that the dissipative contribution originating from the gradient $(\partial u/\partial z)$ is inversely proportional to θ_o whereas that corresponding to $(\partial u/\partial x)$ bears a linear relationship. In practice, since θ_o is small (say $\leqslant 1$ radian) the latter contribution to $T\dot{S}$ amounts only to about 5% or less of the total and may therefore be reasonably neglected [15].

Energy Balance of Motion

Having established the form of the work (per second), Uf, provided by the capillary force, f, and the energy dissipated by viscous drag, $T\dot{S}$, we may set the two equal to obtain a dynamic energy balance :

$$U\ \pi\ \gamma\ \theta_o^2\ 1^{-1}\ \delta\ (t) \sim 6\ \pi\ \mu\ U^2\ r_o\ L/\theta_o \tag{26}.$$

Since the overall speed of motion of the drop, U, is equivalent to the rate of disappearance of the spur by engulfment of the heterogeneity $(-d\delta/dt)$, equation (26) may be simplified to :

$$\frac{d\delta(t)}{dt} \sim \frac{-\ \gamma\ \theta_o^3\ \delta(t)}{6\ \mu\ r_o\ 1\ L} \tag{27},$$

which has the solution :

$$\delta(t) = \delta(0)\ \exp\ (-t/\tau) \tag{28},$$

where the time constant, τ, is given by :

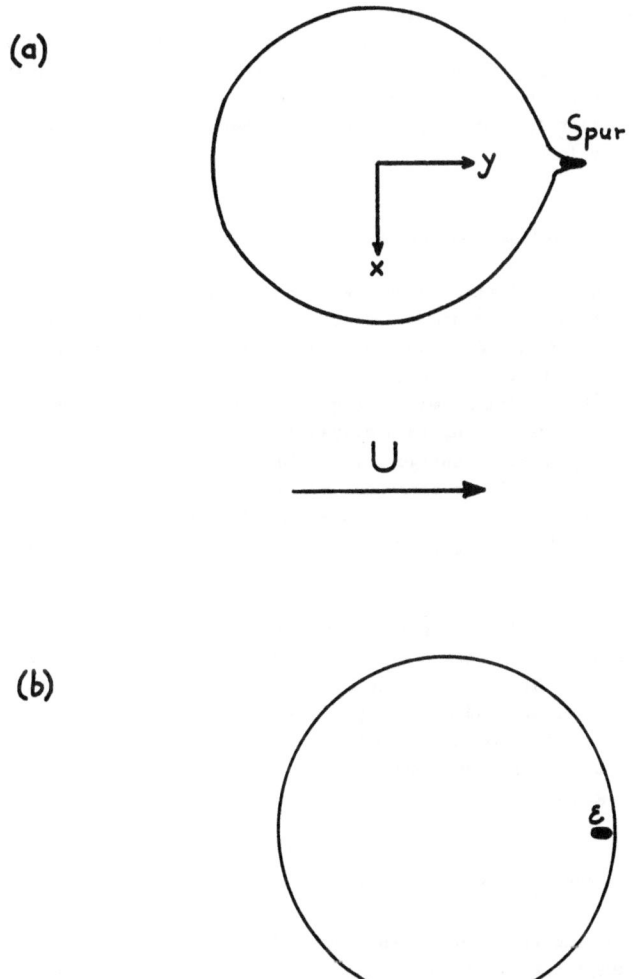

Figure 3 . (a) Plan view of a spur on the drop periphery about to become engulfed (b) after motion of the drop in direction y. Disappearance of the spur is an exponentially decaying process.

$$\tau = \frac{6 \mu r_o 1 L}{\gamma \theta_o^3} \qquad (29).$$

We thus arrive at the conclusion that the kinetics of engulfment of the surface heterogeneity obeys an essentially exponential law with a time constant directly related to the characteristic capillary speed V^* ($= \gamma/\mu$). The process whereby the drop moves towards its state of thermodynamic equilibrium absorbing the pheripheral heterogeneity is shown schematically in Fig.3.

DISCUSSION AND CONCLUSIONS

Wetting hysteresis is a phenomenon generally more commonly observed with liquids of low surface tension and on solids giving rise to low intrinsic contact angles, θ_o . Typically, one expects drops of an organic liquid resting on a high energy surface, such as a metal, for example, to exhibit departure from axial symmetry frequently. In contrast, drops of mercury on glass, corresponding to a system of high γ and θ_o, generally conserve their axisymmetric configurations. This propensity for marked hysteresis with low γ and θ_o is borne out by equations (8) and (9) which show an inverse relationship between triple line deformation, or contact angle modification, and both liquid surface tension and intrinsic contact angle.

Distortions of the meniscus resulting from peripheral heterogeneities on the solid, although more marked near the triple line, exist up to the centre of the drop. This fact helps elucidate drop behaviour on a composite surface. Cassie and Baxter [17] showed that on a piecewise continuous heterogeneous solid surface consisting of two types of material, a and b, giving rise respectively to intrinsic contact angles θ_a and θ_b for a given liquid, the measured contact angle, θ_m, should be given by :

$$\cos \theta_m = f^* \cos \theta_a + (1 - f^*) \cos \theta_b \qquad (30),$$

where f^* represents the fraction of the composite surface occupied by a. Nevertheless, except for the singular case of the triple line falling on an a/b frontier when θ may take on any value between θ_a and θ_b (the so-called phenomenon of *canthotaxis* [18]), the wetting front will come to rest either on a or on b and thus the contact angle will be either θ_a or θ_b but not θ_m. This is clear from consideration of the local equilibrium governed by Young's equation. Likewise, a simple variational approach predicts that the drop profile should correspond to a spherical segment (in the absence of gravity) of contact angle θ_m . The paradox arises because the variational argument and that leading to equation (30) implicitly assume that the dimensional scale of the heterogeneous patches is small compared to overall drop size. The present treatment is more specific in that it does not "average out" heterogeneities. It can be seen that the overall meniscus corresponds essentially to a spherical segment but that inhomogeneities distort the shape. This effect is mainly local but exists also to a lesser extent farther afield.

Equation (9) shows how Young's equation is obeyed locally even for the case when heterogeneities are present (provided the drop is at equilibrium with $A_1 = B_1 = 0$) since :

$$\cos (\theta_o - \Delta\theta) \sim \cos \theta_o + \theta_o \, \Delta\theta \sim (\gamma_{SV} - \gamma_{SL} + \epsilon)/\gamma \sim \cos \theta_o + \epsilon/\gamma \qquad (31).$$

Thus again θ_m is shown to correspond to an average value not taking into account the local behaviour near peripheral heterogeneities. The validity of equation (30) on a large scale is, of course, recognised but it does not allow for fine structure on the meniscus and triple line.

Equation (8) predicts a continuous triple line forming the drop contour since the Fourier series has terms of O (n^{-2}) (both A_n and B_n are of O (n^{-1})). This result, which may be expected, is confirmed by the logarithmic representation of equation (16). Nevertheless equation (9) allows for discrete jumps in θ. These can clearly not exist in reality and the mathematical cause is that $\Delta\theta$ is evaluated to first order and therefore along radius r. Effects from η_ϕ appear only in second order, mathematically intractable terms. In the context of this study, we may consider the contact angle actually *on* a frontier between zones of different surface free energy to be mathematically indeterminate, although it must clearly lie within the range defined by the intrinsic values for the liquid on the two neighbouring areas [18]. In a physically real situation, of course, the transition between zones will be governed by molecular structure. Treatment of this is beyond the scope of the present continuum approach.

As shown above when the Fourier coefficients A_1 and/or B_1 of equation (5) are non-zero, the drop is not strictly at equilibrium. For simple cases, such as that of the single heterogeneity presented here, a suitable choice of origin for the polar angle Φ leads to A_1 being zero but B_1 finite. The exponentially decaying movement of the drop due the term B_1, or alternatively the force f, has a time constant, τ, given by equation (29). Typically for a small drop of an organic liquid, we may expect the following approximate values : $\gamma \sim 30$ dyn.cm^{-1}, $\mu \sim 10^{-1}$ P and $r_o \sim 1$ mm. An intrinsic contact angle, θ_o, of ca.0.2 rad. and a value for L of 4 are not unreasonable. Taking $w \sim 10$ μm, we obtain a value for the time constant τ of ca.5s. Assuming the drop to be placed on the solid surface and spreading to its approximately equilibrated shape defined by the intrinsic contact angle, θ_o, before the lateral movement caused by f ensues (the two steps will be to some extent concomitant but it is convenient to separate them in a first approximation), it has previously been shown [19] that the equivalent time constant is given by $\tau_a = r_o L/(V^{\bullet}\theta_o^3)$. For the values given, this corresponds to $\tau_a \sim 0.17$ s. The time constant for "crawling" is thus about 30 times greater showing that this process is far slower than the initial spreading. This may be expected intuitively since lateral movement is controlled by weak fluctuations to the mean surface free energy of the solid whereas it is the average value of γ_{sv} (together with γ_{sL} and γ) which determine the basic spherical segment shape.

Although not dealt with here, motion of the drop may bring it into contact with other surface heterogeneities which in their turn, may "pull" the liquid mass towards them (or repulse it if ϵ_o is negative). This situation is somewhat analogous to the erratic movements of rain drops on a window pane. When a rain drop descends a vertical window, provided its mass is not too great (in which case gravity overrules capillary effects completely and the drop runs straight down), sporadic changes of both speed and direction may be observed superimposed on the general downwards tendency. These are due to a large extent to local heterogeneities of the glass surface (dirt, grease spots etc.) provoking local capillary movement similar in principle to that analysed above for the simplest case of a single peripheral inhomogeneity. The general treatment would, of course, be quite complex.

Lateral movement of the drop leading to the engulfment of heterogeneities is reminiscent of the process of phagocytosis observed in microbiology. When a cell, such as a macrophage encounters an "impurity" such as a bacterium, the former embarks upon a process of ingestion of the latter [20] not dissimilar to the engulfment of free energy imperfections on a solid by a liquid drop. Clearly a living cell has a very complex structure unlike a simple drop of liquid and so the similarity of the process may be only apparent. Nevertheless, it is felt that at least the onset of phagocytosis must be controlled by surface energetics [21].

In this study, it has been implicity assumed that surface heterogeneity is due to energetic variations caused by such factors as the presence of different chemical species or local specific adsorption on the solid. Nevertheless, the approach may be used to treat physical inhomogeneity (roughness) provided the extent is limited [14]. In particular changes in surface topography must not be too abrupt. This may prove to be of use for considering the effects of wetting hysteresis on machined surfaces.

We have treated various static and dynamic aspects of capillary phenomena on slightly inhomogeneous solids in this study. Although various conditions have been imposed in order to render the mathematics tractable, the basic physics should apply to more complicated systems, such as those of higher intrinsic contact angle and/or those in the presence of external forces fields such as gravity.

ACKNOWLEDGEMENT

Although this study is recent, its origins lie in an interest in wetting phenomena aroused many years ago by the late Dr. W.C. Wake when the author was a post-graduate student. It is out of friendship and respect for this doyen of adhesion science that the present paper is dedicated to his memory.

REFERENCES

1. Hansen, R.S., Miotto, M., J. Am. Chem. Soc., 1957, 79, p. 1765.

2. Johnson, R.E., Dettre, R.H., Adv. Chem. Ser., 1964, 43, p. 112.

3. Eick, J.D., Good, R.J., Neumann, A.W., J. Colloid Interface Sci., 1975, 53, p. 235.

4. Good, R.J., J. Colloid Interface Sci., 1977, 59, p. 398.

5. Yasuda, H., Sharma, A.K., Yasuda, T., J. Polym. Sci., Polym. Phys., 1981, 19, p. 1285.

6. Carré, A., Moll, S., Schultz, J., Shanahan, M.E.R., Adhesion 11, ed. Allen, K.W., Elsevier Applied Science Publishers, London, 1987, ch. 6.

7. Shanahan, M.E.R., Adhesion 14, ed. Allen K.W., Elsevier Applied Science Publishers, London, 1990, ch. 5.

8. Laplace, P.S., Mécanique Céleste, Suppl. au X livre, Coureier, Paris, 1805.

9. Shanahan, M.E.R., J. Chem. Soc., Faraday Trans. 1, 1984, **80**, p. 37.

10. Shanahan, M.E.R., J. Phys. D., 1989, **22**, p. 1128.

11. Smirnov, V.I., A Course of Higher Mathematics, trans. Brown, D.E., Pergamon, Oxford, 3rd. edn., 1964, Vol. IV, p. 201.

12. Courant, R., Hilbert, D., Methods of Mathematical Physics, Wiley, New York, 1953, Vol. I, p. 179.

13. Massey, H.S.W., Kestelman, H., Ancillary Mathematics, Pitman, London, 1964, p. 310.

14. Joanny, J.F., de Gennes, P.G., J. Chem. Phys., 1984, **81**, p. 552.

15. Shanahan, M.E.R., J. Phys. D., 1990, **23**, p. 321.

16. de Gennes, P.G., Rev. Mod. Phys., 1985, **57**, p. 827.

17. Cassie, A.B.D., Baxter, S., Trans. Faraday Soc., 1944, **40**, p. 546.

18. Haynes, J.M., Materials Science in Space, ed. Feuerbacher, B., Springer Verlag, 1986, p. 129.

19. de Gennes, P.G., C.R. Acad. Sci., Paris, 1984, **298(II)**, p. 111.

20. Kaplan, G., Scand J. Immunol., 1977, **6**, p. 797.

21. Neumann, A.W., Advances Coll. Interf. Sci., 1974, **4**, p. 105.

9

THE USE OF OPTICALLY STIMULATED ELECTRON EMISSION FOR THE DETECTION OF SURFACE CONTAMINATION

C J ALLEN, C KERR and P WALKER
SCT Materials, AWE, Aldermaston, Berkshire

INTRODUCTION

When metals, semiconductors, and some polymers are exposed to high energy radiation such as ultra-violet, some electrons at the surface will absorb the photon energy and escape from the surface. This phenomenon is known as optically stimulated electron emission (OSEE). At a metal boundary surface the potential energy of an electron approaches zero and thus functions as a barrier to the escape of electrons. In order to escape from the surface an electron requires additional energy over that associated with its velocity component perpendicular to the barrier surface. This energy may be provided by the processes of thermionic or field emission, by secondary emission from primary electrons, ions, atoms or photons [1]. In the process described, the energy is provided by UV radiation and the emitted electrons are collected across an air gap by a biased collector and measured as a current. Not all the escaping electrons will be collected, some will be deflected back to the surface, combine with oxygen or diffuse as free electrons. The typical current level is 10^{-10} - 10^{-12} amperes [2]. Any surface layer present, depending on its own photoemission characteristics will either enhance or attenuate this current.

These changes in current can be used to detect the presence, position and thickness of a surface layer whether organic, oxide or other inorganic, and makes the OSEE technique particularly valuable for assessing the degree of contamination of a surface. Relatively non-emitting materials, and therefore attenuating, include hydrocarbon oils, greases and silicones, which are all common contaminants harmful to the attainment of strong adhesive bonds, good adhesion of surface coatings and the achievement of effective pretreatment.

The fact that a contaminant may be emitting rather than attenuating does not necessarily rule out the possibility of using OSEE, as even emitters may change the strength of the measured signal. Metal powders and fragments can be located by their enhanced emittance over a low emittance background.

Equipment

The equipment described is the Model OP1010 Surface Quality Monitor
manufactured by Photo Acoustic Technology Inc. of Newbury Park, California
and is known as PATSCAN. The equipment is illustrated in Figure 1 and
consists of an IBM PC Computer with high resolution monochrome monitor
with high resolution graphics capability (A), a control unit (B), a
sensor with vertical height adjustment (c), and a 150 x 150mm scanning
stage with NEAT Controllers (D).

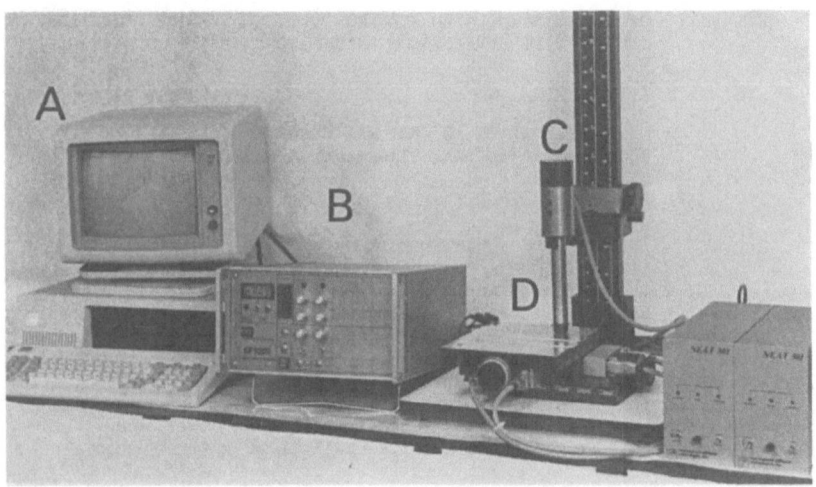

Figure 1. PATSCAN Model OP1010 Surface Quality Monitor.

The scanning stage may be commanded to control the direction, speed
and scanning interval in the X and Y direction and may be halted or
reversed as required. The sensor(s) contain an ultra-voilet lamp, a
detector, detector biased circuitry and amplification electronics. The
aperture size can be altered with suitable fitments to reduce the area
inspected. A rotary switch enables the amplifier gain to be set as
required and signal lights indicate if readings are within a pre-set
window.

The control unit houses the UV lamp power supply, DC power supply,
the electronic circuit boards and various displays.

The PC/monitor containing dedicated PATSCAN software enables the
user to control the motion of the scanning stage in terms of speed, area
covered, number of data points etc. In the graphic display mode a three
dimensional image of the data collected may be manipulated as required.
The X and Y positions, axis, speed etc are on continuous display.

FACTORS AFFECTING EQUIPMENT PERFORMANCE

There are several factors known to influence the use of PATSCAN, the most
important of which are discussed below.

Sensor Distance

The sensor output is markedly dependent on the distance of the sensor
from the test substrate, and three factors are important [3], (i) as the
sensor distance increases, the UV intensity decreases, and fewer
secondary electrons are released, (ii) electron losses due to collision
increase with distance, and (iii) collection efficiency decreases due to
a decrease in the electric field strength. The effect of these
accumulative effects on sensor readings is shown in Figure 2 for three
aperture sizes on a PATSCAN type probe. Gause [3] suggests a maximum
distance of 6.35mm while data from Smith [4] indicates a distance of
1.2mms for the same percentage current drop using an unspecified probe.
Clearly this is a parameter to be established for individual probe and
aperture settings.

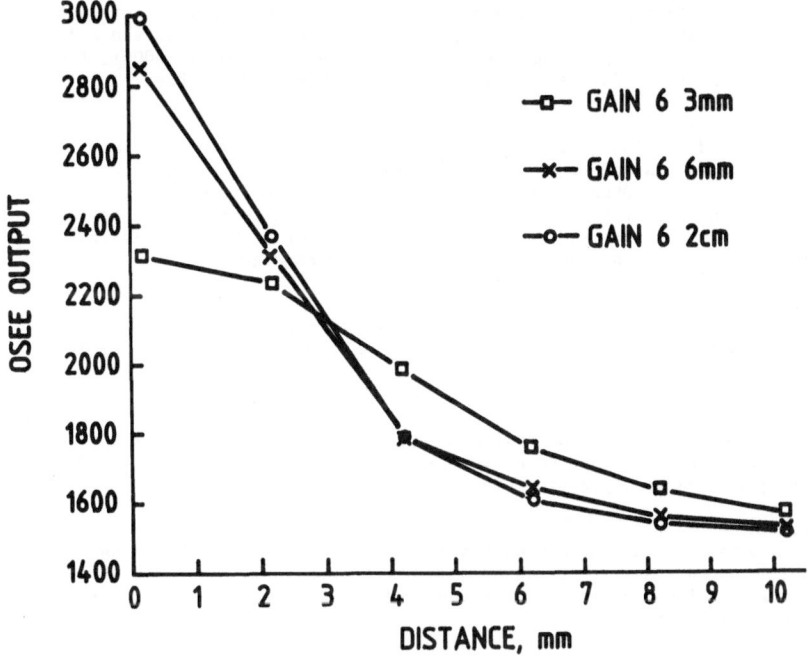

Figure 2. Effect of sensor distance on OSEE signal.

Scan Time

Scan times ie duration of exposure of the test surface to UV light can be
important as the peak current reading may fall with even short time

exposure due to electron stripping of the surface layer, most noticeable on some organic emitters.

On a contaminated surface, the measured OSEE may increase with exposure time due to the disassociation of the organic molecules under intense UV irradiation [5]. In practice this variable is eliminated by the recording of the peak OSEE response and a constant short exposure time.

Scan Interval
The scan interval ie distance between measured areas is only important in terms of the completeness of the examination. Large intervals require the use of a large aperture probe and small intervals a small aperture probe. For identification of local areas of contamination, small scan intervals are recommended if a detailed picture is to be obtained. The equipment described is capable of adjustment over the range 0.03 to 20mm.

UV Wavelength and Intensity
Short wavelength radiation produces higher energy electrons than long wavelength, and the former are more readily collected by the sensor. The rate of electron emission from a surface is dependent on lamp intensity which is governed by the current passed. The probes used contain a low pressure mercury vapour lamp emitting over the range 185 to 600nm, the current is preset for maximum lamp life and is not adjustable.

Detection Limits
The lower detection limit for the measurement of contaminant layer thickness is claimed to be 1-2Å [2]. There is effectively no upper limit, as even when the OSEE signal is extinguished by the presence of a thick contaminant layer, the presence of the contaminant is readily detected by cleaning the surface and remeasuring. A layer of silicone resin (RTV 102) 240Å thick has been shown to extinguish the OSEE signal [4].

Operating Modes
The equipment may be used in either of two operating modes, the choice of which is dependent on the particular application to which the equipment is being put.

Peak Detect Mode
This mode is used when establishing upper and lower settings if the equipment is to be used on an accept/reject basis, as for example when being used as an inspection tool on surface having unknown characteristics.

Scan Mode
This mode is recommended when characterising surfaces, identifying contaminated areas, or being used as a research tool. It can also be used for inspection purposes in this mode when large areas or moving surfaces are being investigated.

RECORDING OF DATA

Data may be recorded in three ways:
(a) numerical hard copy (tabular data)
(b) plotted graphic data (also displayed)
(c) normalised plotted graphic data (also displayed)

(a) Numerical Data: In this case block numerical data on the current
measurements is provided. The presence of contamination arising from a
fingerprint on a clean aluminium surface is shown in Table 1. The scan
area is 15 x 12mm and the scan step 0.15mm. The lowest OSEE signal
values coincide with the area of maximum pressure and therefore
contamination level.

TABLE 1
Numerical printout for a fingerprint on clean aluminium

2999	2999	2999	2999	2999	2805	2999	2999
2999	2999	2999	2587	2237	2117	2810	2999
2999	2999	2328	2122	1961	1931	2299	2999
2999	2999	1986	1901	1827	1862	2694	2999
2999	2156	1755	1710	1693	1811	2784	2999
2999	2230	1667	1649	1653	1834	2784	2999
2999	2460	2056	2188	1653	2522	2810	2999
2999	2999	2350	2463	2410	2622	2999	2999
2999	2999	2999	2999	2999	2999	2999	2999
2999	2999	2999	2999	2999	2999	2999	2999

(b) Graphic Data: Figure 3 is an example of a three-d print-out which
shows a scan of a heavily fingermarked aluminium surface, the scan area
is 150 x 120mm and the scan step 1.5mm. The area of contamination by the
fingers shows as a series of apparent depressions reflecting the lower
OSEE signal values from the contaminated areas.

(c) Normalised Graphic Data: In this case the OSEE signal data from the
original uncontaminated surface is subtracted from that of the
contaminated surface, point by point, before printing. Figure 4 shows
the data from the same panel as Figure 3 but treated in this manner. The
effect is to show the contamination as a series of raised areas more
closely representative of the actual situation.

This latter approach to data record requires data for a clean substrate
to be obtained at the same scan interval and gain setting before
contamination, and is therefore not universally possible. Methods (b)
and (c) are the most useful allowing the contaminated areas to be located
and accurately pin-pointed. The choice between (b) and (c) must be a
personal one and the present authors have different preferences.

DISPLAY OPTIONS FOR GRAPHICS

Graphic data can be displayed and printed in several forms. The
elevation angle ie the angle of tilt from the horizontal can be changed
over the range 0-90°, the quadrant, ie the side of the three dimensional
graph viewed, can be selected and the vertical height can be enhanced. A
zoom option is also available for selecting and magnifying small areas.

The visual effect of changing the quadrant viewed is shown in Figures
5 and 6 which show Quadrants 1 and 4 of a heavily contaminated chromic
acid etched aluminium panel, the data is normalised.

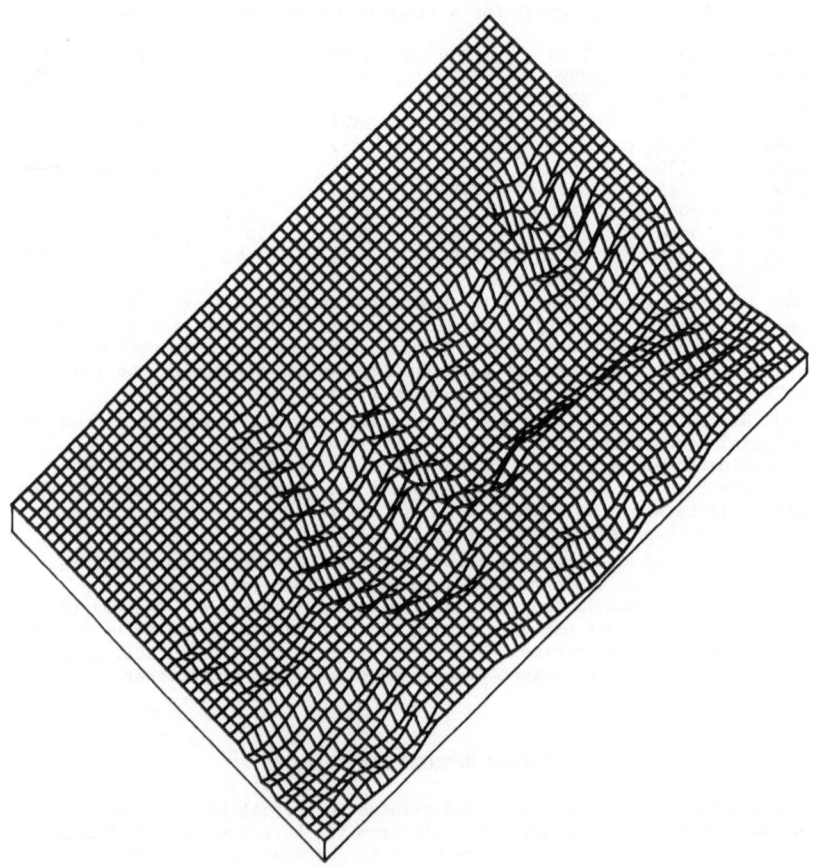

Figure 3. 3D Print-out of a heavily fingerprinted surface.

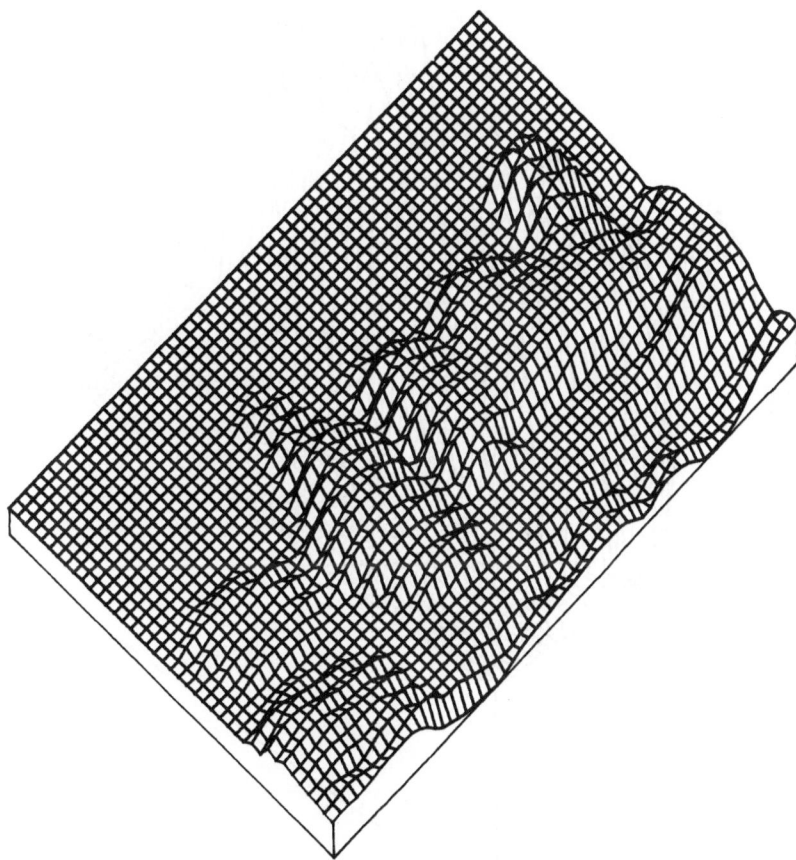

Figure 4. Normalised graphic print-out of a heavily fingerprinted
surface.

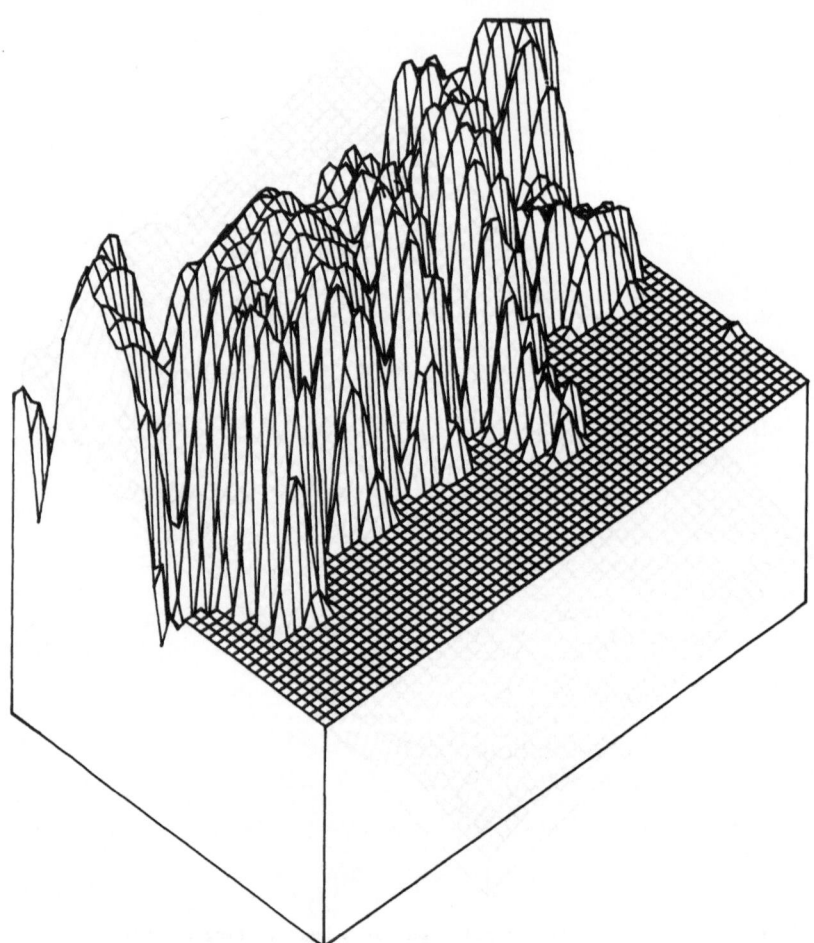

Figure 5. Quadrant 1 normalised print-out.

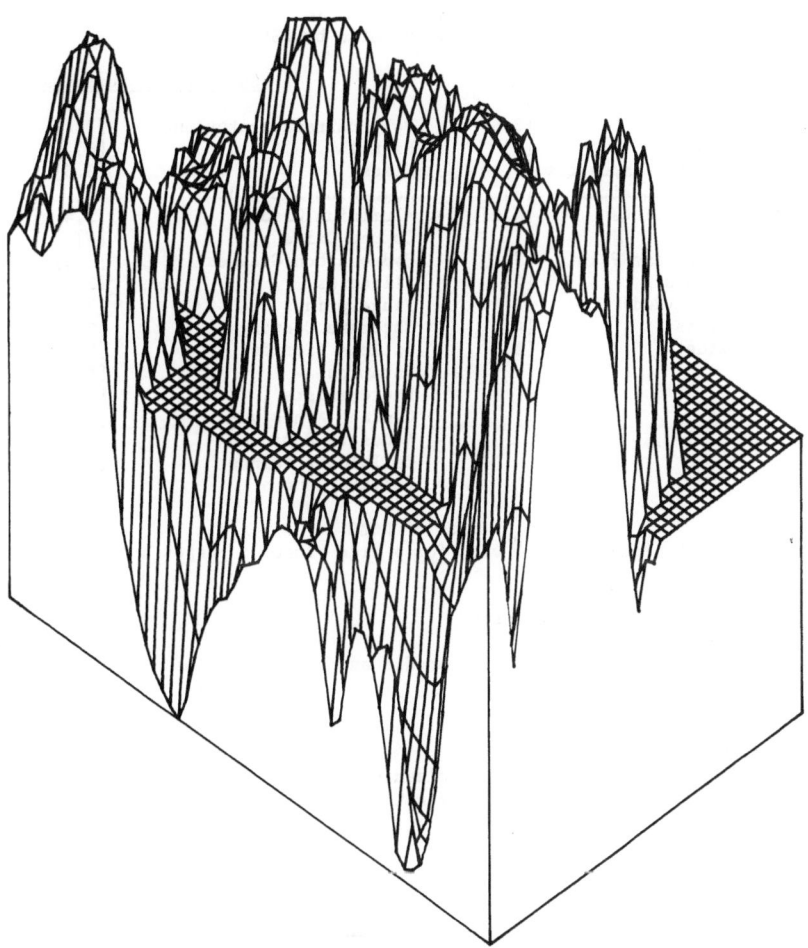

Figure 6. Quadrant 4 normalised print out.

Surface Investigated by OSEE

Sensor response depends on the photoelectron work function of the material relative to the maximum usable UV energy reaching the surface. If a material has a work function less than 5 eV it should produce a measurable photocurrent [3]. Surfaces investigated by other workers are shown in Table 2.

TABLE 2
Surfaces investigated by OSEE

Emitting Surfaces		Non-emitting Surfaces	
Most Metals	[3,4]	Poly tetrafluorethylene	[3]
Semiconductors	[5]	Nickel	[6]
Graphite/Epoxy Composites	[3,4]	Glass	[3.5]
Glass/Epoxy Composites	[3]	Magnesium Fluoride	[3]
Some Epoxide Paints	[4,5]	Aluminium Oxide	[3,4,6]
		Copper Oxide	[4]
Some Polyurethane Paints	[5]	Hydrocarbon Greases	[2,5]
Magnetic Discs	[2]	Silicones	[2,3,4,5]
Pencil Marks (Red)	[2]	Fingerprints	[3]
Photo resists	[4]	Chromate Coatings	[3]
Nickel Oxide	[6]	Silicon Dioxide	[2,4]
		Ceramics	[5]
		Photoresists	[2]
		Contact Adhesives	[4]
		Adhesive Residues	[4]
		Cellulosic Paints	[5]
		Polyethylene	[5]

Uses of OSEE

The PATSCAN equipment may be used for a wide range of purposes including:

Inspection:
efficiency of cleaning processes
presence of contamination
uniformity and quality control of anodic and chromate films
detection of flaws.

Research:
detection of contamination
identifying suitable solvents for cleaning
correlation of bond strength with cleanliness of surface
investigations of adhesion failures
identification of sites of failure

Practical Investigations Using OSEE

Detection of Contaminants

Figure 7 shows the graphic data (not normalised) for a carbon fibre reinforced plastic composite deliberately contaminated with silicone fluid, hydraulic oil and orange juice. The area scanned is approximately 120 x 20mm and the scan step 2.0mm, the position and extent of each of the three contaminants can be seen clearly. The maximum and minimum OSEE signal values were 1646 and 995.

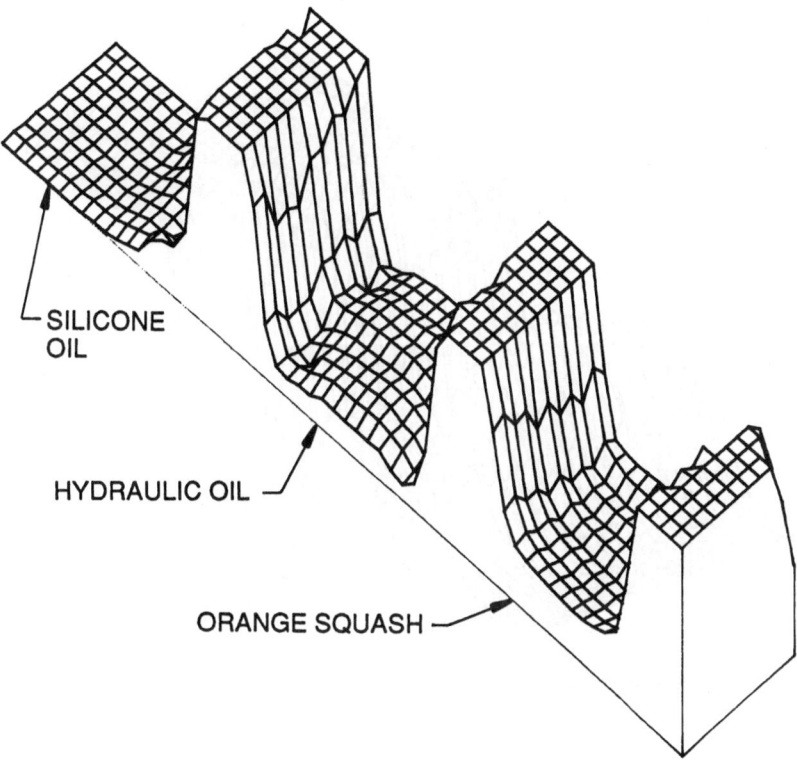

SILICONE
OIL

HYDRAULIC OIL

ORANGE SQUASH

Figure 7   Non-normalised print-out of a contaminated CFRP surface

The graphic data for an abraded degreased panel deliberately
contaminated with silicone grease and a PTFE spray is shown in Figure 8.
Once again the data is not normalised and the scan area is 90 x 60mm and
the scan step 1.5mm. Four distinct areas can be identified, the PTFE
sprayed area on the extreme left of the panel, on immediately adjacent
area showing a decreasing contamination level, a silicone grease
contaminated area and the clean panel on the right: The interesting
feature here is that the silicone contaminated area shows a sharp
demarcation between the contaminated and non-contaminated areas, the PTFE
spray does not. This indicates the dangers attendant on the use of
release agents applied by spray as they may contaminate large areas. The
maximum and minimum OSEE signals were 1726 and 1499.

Figure 9 shows normalised graphic data for a clean aluminium panel
deliberately contaminated with stripes of a mould release agent Releasil
7 and molybdenum disulphide. The scan area is 90 x 60mm and the scan
step 1.5mm, and the maximum and minimum OSEE signals 1492 and 905
respectively. Both contaminants were applied by finger and the graphics
indicate that the Mo S_2 was in the form of a uniform thickness non-

spreading layer, the Releasil 7 showed signs of spreading and non-uniform film thickness.

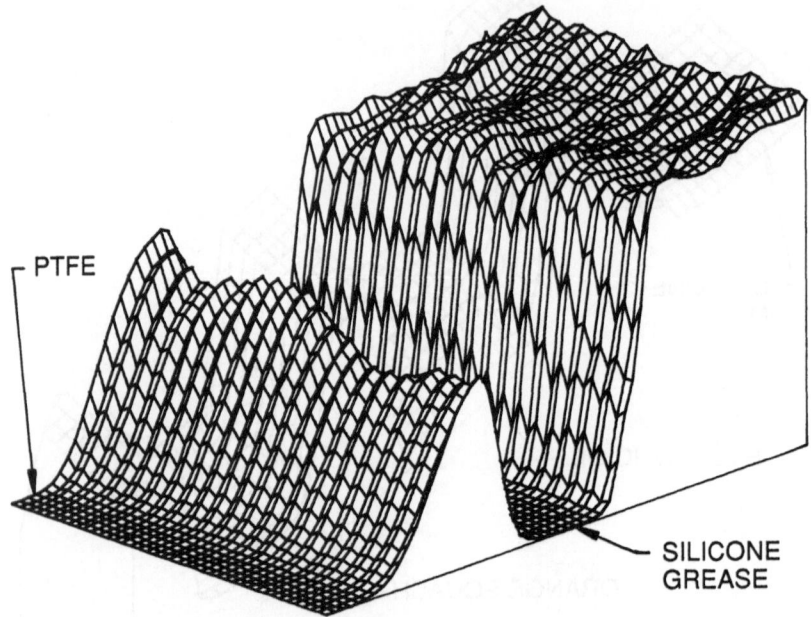

Figure 8   Non-normalised printout of a contaminated aluminium surface

Efficiency of Solvent Cleaning.

Figure 10 shows the non-normalised graphic data for a clean abraded aluminium panel which has been contaminated with a silicone oil and subsequently solvent cleaned. Five distinct areas can be identified. On the extreme left the clean surface can be seen with a slight trace of contaminant in the centre, the area of contamination marked "oil", followed by 1 and 2 wipes and finally spray; the solvent used was acetone. it can be seen that it is not until the panel has been given three successive solvent treatments that most of the contaminating silicone oil has been removed.

The efficiency of solvent cleaning processes using toluene for the removal of hydrocarbon oil contamination from a mild steel panel is shown in Table 3. The separate panels were given the accumulative treatments shown, thus panel 2 was given one solvent wipe and panel 6 given the treatments for 2-5 inclusive plus vapour degrease. The OSEE signal output increased with increasing level of cleaning and the bond strength of a two pack epoxide adhesive as determined by the torque shear method, [7] also increased with improved cleaning.

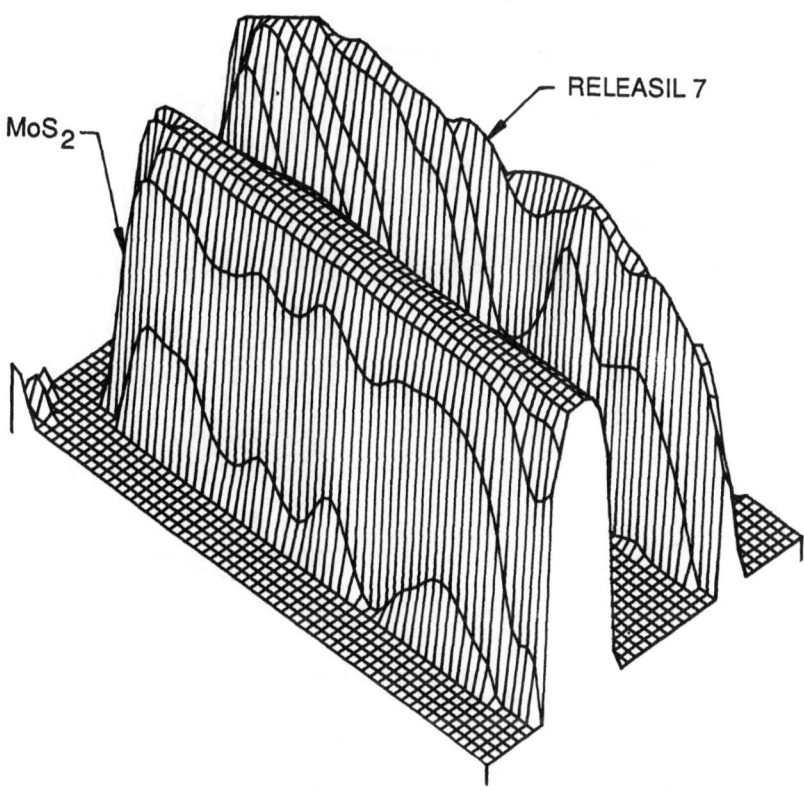

Figure 9 Normalised print-out of an MoS$_2$ and Reasil contaminated panel

TABLE 3
Effect of solvent cleaning on OSEE signal and bond strength

Cleaning Process		OESS Signal Minimum	OESS Signal Maximum	Bond Strength MPa
As received	(1)	1512	1521	Zero
1 Solvent Wipe	(2)	1511	1575	8.7
2 Solvent Wipes	(3)	1557	1637	12.9
3 Solvent Wipes	(4)	1584	1704	16.8
Solvent Spray	(5)	1676	1758	19.3
Vapour Degrease/ Solvent Spray	(6)	1696	1924	22.7

The scatter of results was extremely high and it would have been interesting to correlate the local OSEE signal with the measured individual bond strength readings.

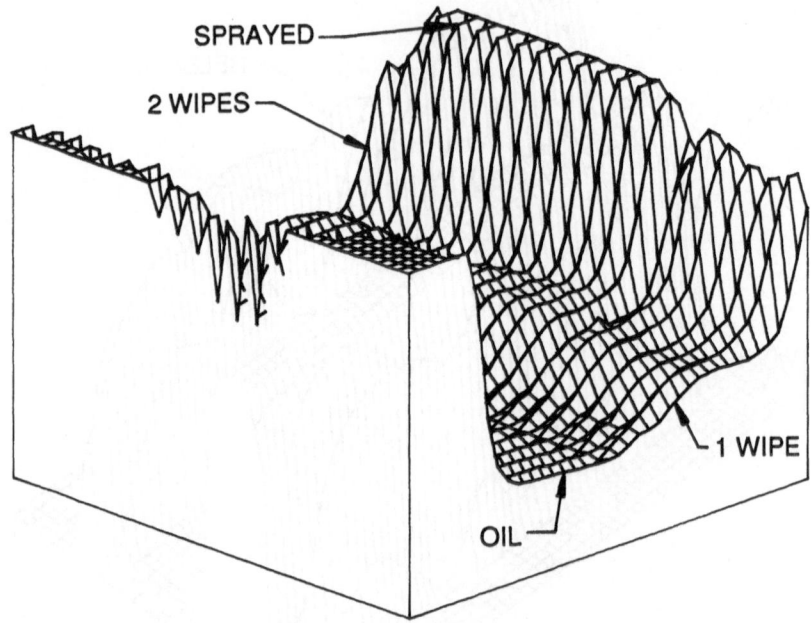

Figure 10   Effect of solvent cleaning on the OSEE output from a silicone
            oil contaminated panel

Ageing Of Prepared Surfaces

Changes in a prepared surface with time may be important when adhesively
bonding or coating materials which grow an oxide film or suffer other
changes such as migration of low molecular weight species to the surface
or collect contamination by static attraction. The change in the OSEE
signal of an abraded aluminium panel with time is shown in Figure 11,
where the OSEE output is plotted against time of exposure, the maximum and
minimum outputs are shown.

The OSEE signal diminishes rapidly over the first twentyfour hours as the
thickness of the non-emitting oxide film increases. There is little
further change in output after fifty hours ageing.

Detection of Scratches

The graphic data for a scratched aluminium panel is shown in Figure 12
where the scratches have penetrated through the non-emitting surface oxide
layer resulting in an enhanced OSEE signal output from the scratches. The
scan area is 90 x 60mm and the scan step 1.5mm.

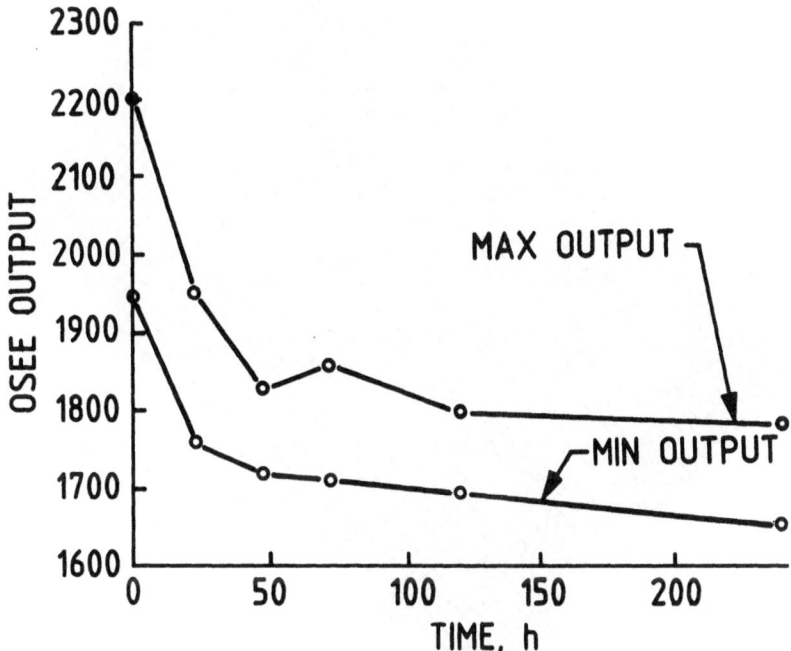

Figure 11. Effect of ageing on the OSEE output from an abraded aluminium panel.

CONCLUSIONS

It has been demonstrated that OSEE is a useful tool for the detection of contamination on substrates prior to bonding, with the capability of being able to scan the entire area, OSEE also provides information on the distribution of contaminants over the surface and is capable of producing both graphic and numerical data. The technique has shown that solvent cleaning processes can be inefficient resulting in a distribution of contamination over an area rather than its complete removal. The method should provide a useful tool in the solution of problems involving contamination of substrates.

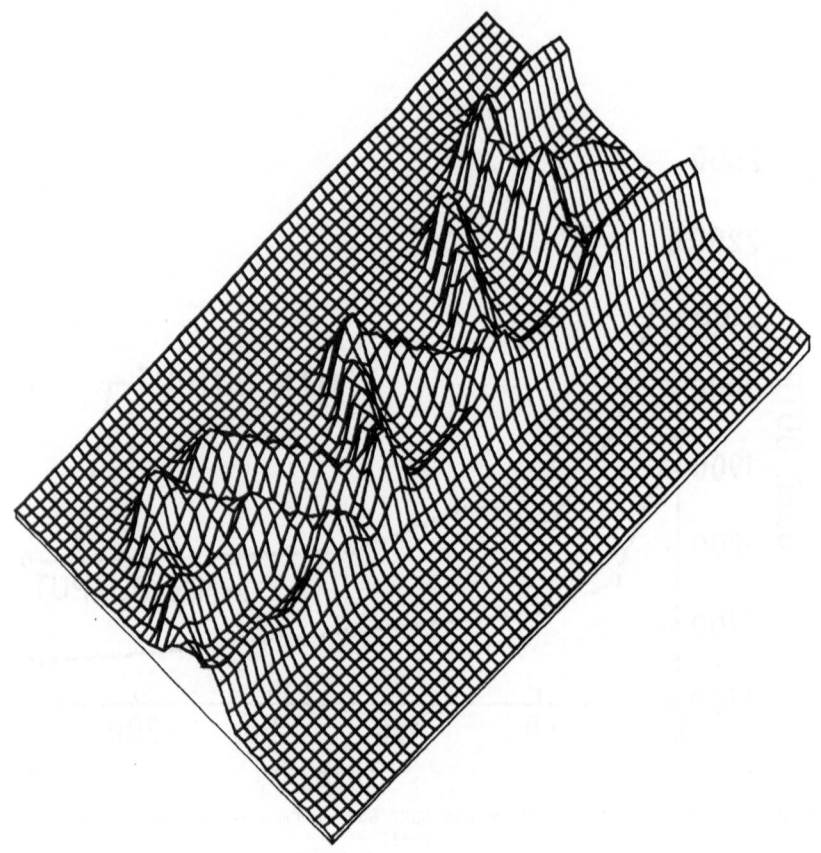

Figure 12   Detection of scratches

REFERENCES

[1] Weissler, G L, Photoionization in gases. In <u>Encyclopaedia of Physics</u>, ed, Flugge, Springer - Verlag, Berlin, 1956, vol 21, p 342.

[2] Arora., A., Surface Contamination Measurement and Control By Nondestructive Techniques, 21st Annual Technical Meeting of the Institute of Environmental Sciences, Florida, April-May 1985.

[3] Gause, R. L., A Noncontacting Scanning Photoelectron Emission Technique For Bonding Surface Cleanliness Inspection. Fifth Annual NASA NDE Workshop, Florida, December 1987.

[4] Smith, T., Surface Quality Unit For Inspection by Nondestructive Testing (SQUINT) with Photoelectron Emission SAMPE Quarterly 15, 2, p 6-13, January 1987.

[5] Operating Manual, OMOP101A Surface Quality Monitor, Photo Acoustic Technology Inc, December 1986.

[6] Smith, T., Photoelectron Emission From Aluminium And Measured Nickel In Air, Journal of Applied Physics 1875, 46, pp 1553-1558.

[7] Holloway, M. W. and Walker, P., A Method For The Measurement Of The Adhesion Of Surface Coatings Under Service Conditions. Journ, Oil & Col, Chem, Assoc, 1964, 47, pp 112-130.

10

XPS STUDIES OF THE INTERFACE BETWEEN PMMA AND ALUMINIUM

WULFF POSSART

Institute of Polymer Chemistry "Erich Correns", Academy of
Sciences of the GDR, Kanstrasse 55, Teltow, DDR-1530 GDR

WOLFGANG UNGER

Central Institute of Physical Chemistry, Academy of Sciences
of the GDR, Rudower Chaussee 5, DDR-1199 Berlin, GDR

1.   INTRODUCTION

For many years x-ray photoelecton spectroscopy (XPS) has been
successfully applied to the characterisation of the
interphase between polymers and (oxidised) metal surfaces.
The identification of elements and their relative
concentration in a layer equal to the so called "sampling
depth" of the method appears as a matter of routine nowadays.
Concentration gradients are detected by combining XPS with
ion etching; however this technique gives rise to some
controversy concerning the mixing that the ion beam might
induce in the matrix.

With respect to adhesion this information is useful but of
limited value since it is the interactions between the
polymer and the metal oxide layer that deserve our special
interest.

Hence the problem has to be redefined in such a way that XPS
may answer not only the question of the chemical bonding
state of the elements, but further how this state is modified
by specific interactions inside the interphase.

This problem is a rather complex one, both from the
experimental and the theoretical point of view. The
resolution and the reproducibility of the photoelectron
spectra must be good enough to probe the concentration and
the different bonding states of the elements in the
interphase between polymer and metal as compared with the
conditions in the bulk phases. Hence an advanced technical
apparatus and a well defined preparation technique have to be
combined with a physically sound methods of spectra
decomposition by peak fitting. The discussion of such
methods has by no means finished yet. But in our opinion it
deserves growing attention since the parameters of the XPS
line shape contain the physically relevant information about
the photo process and hence the electron state detected by
it.

Last but not least, we should keep in mind that such an
interpretation of photoelectron spectra is mostly confined to
core levels today. Thus, whenever modifications of the
electronic bonding state are detected and described they are
not immediately related to the adhesive interactions which
occur between the valence levels. As a consequence, XPS
investigations in adhesion science should be complimented by
a quantum chemical modelling of the way an adhesive
interaction causes disturbances of the state of core level
electrons. We note in passing that it seems insufficient to
confine the model to traditional chemical bonds between two
atoms. The results of research in catalysis and adsorbtion
favour interactions between more than two partners, like
those in complexes of clusters.

The number and difficulty of the problems involved with the
question of adhesive interaction are too large to be solved

at once. What we intend with this paper is to describe our
first steps on the way to their solution and to promote the
discussion of these items.

2. EXPERIMENT

Aluminium was evaporated onto the polished surface of silicon
wafers. The subsequent storage under laboratory atmosphere
in an desiciator for a few days was intended to complete the
build up of the natural layer of oxides, hydroxides and
water. The polymer films consist of radically polymerized
poly(methyl methacrylate) (PMMA) possessing a mean molecular
mass of $\overline{M}_n \approx 5 \times 10^4$. After depositing a suitable amount of
polymer-chloroform solution onto the aluminium surface the
film formation proceeded in a closed box under a solvent
saturated atmosphere, thus assuring a slow solidification of
the polymer layers with thicknesses from 2nm to ca 1μm.

The photoelectron spectra were measured in the angle-resolved
mode (ARXPS) with nonmonochromatized Al Kα excitation
employing an AEI ES 200B electron spectrometer linked to a DS
100 data system. The analyzer resolution was set to give an
Au 4f 7/2 peak with a full width at half maximum (FWHM) ΔE of
2.0eV. According to the literature [1,2] sampling depths of
about 8nm and 3nm are estimated for the applied take-off
angles of 20° and 70° respectively. The polymer thickness
values in the nm-range are estimated from the Al/C atomic
ratio [3] supposing the polymer film to be a homogeneous,
flat overlayer.

The investigations included intact samples with nominally 2,
4, 6nm and 1μm PMMA layers as well as the aluminium substrate
for comparison. No other elements than C, 0 and Al appeared
in the spectra. Figure 1 contains examples of three spectra
after removal of background. Further details are discussed
below.

3. RESULTS AND DISCUSSION

The relatively featureles structure of the observed specra confirms the expectation that a qualitative analysis proves to be insufficient for conclusions concerning adhesive interactions. Therefore a decomposition procedure has to be applied. In the preliminary version used here the corresponding software contains four kinds of background removal together with a Marquard fitting procedure (similar to a version provided by Hughes and Sexton [4]) on the basis of a Gauss-Lorentz product as the line shape function. This function supplies the energetic position, the intensity, and the FWHM as the characteristic features of each spectral component. Furthermore the mixing ratio between Gauss and Lorentz functions in the line shape is fitted as a general parameter. The data processing procedes as a graphically-supported computer dialogue.

All the spectra reported in this paper were reduced by a Shirley-background [5] prior to fitting.

Generally it is not possible to decide on the basis of the fitting results alone how many components the spectrum comprises. Some additional physical arguments must be taken into account.

First, the chain ends of the PMMA molecules, with the given molecular mass, possess a root mean square distance of 14nm [6]. Thus, three components for the C 1s spectra and two components in the 0 1s signal should occur at least for the 20^0 take-off angle because the corresponding sampling depth of ca 8nm is of the same dimension as a macromolecule. Figure 1 depicts the structural element of PMMA and gives the connection between the spectral components and the atomic bonding states discriminated by the fitting. The atomic symbols underlined in Fig 1, and in the following text indicate that element to which the spectral feature considered refers. Notice that the given experimental

152

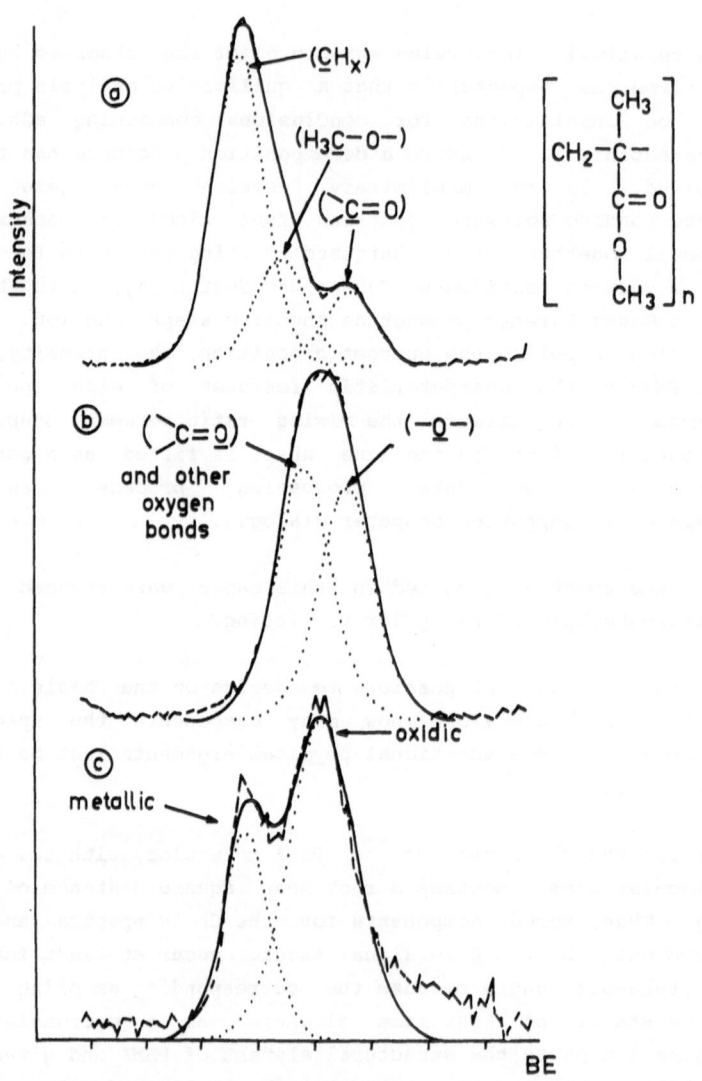

Fig 1: Three examples of XPS from PMMA on Al substrate.
--- = experimental spectrum after background removal;
.... = fitted components; ——— = fitted spectrum.
(a) C 1s spectrum (repeating chain unit for
convenience); (b) O 1s spectrum; (c) Al 2p spectrum

conditions confine a decomposition of the oxygen spectra to the separation of oxygen belonging to the ether bridges. The remaining bonding states of 0 cannot be resolved due to nearly equal chemical shifts being typical for these bonding states of oxygen. Aluminium should occur as metal and as oxide/hydroxide in the Al 2p spectra.

Secondly we remember the FWHM to be related to the transition probability of the photoelectron. It seems reasonable to suppose a transition probability for the polymer core level electrons that is approximately independent of the distance between atom and interface, since the adhesive interactions should influence the core electrons rather indirectly only. Hence the FWHN for the C 1s and 0 1s components repectively are coupled during the fitting to provide equal values for either element. The aluminium components require a somewhat modified presumption. Due to the great distinction between the metallic and the oxidic bonding state, the FWHM are not set to be equal here but to retain a constant difference during the fitting. This seems reasonable too in the light of Figure 1c.

Finally the mathematical rather than physical sense of the Gauss-Lorentz mixing ratio has to be stressed. According to preliminary fits of the spectra with unconstrained parameters this ratio proved to be nearly the same for all spectra of a certain core level. Hence the mixing ratio was set constant for the actual fitting procedures. The quality of the overall fits of the spectra is convincing under these suppositions - Figure 1

The resulting line parameters for the three elements are shown in Figures 2 - 4 with the abscissa directed towards increasing vicinity between the layer that supplies the photoelectrons and the phase boundary between PMMA and the substrate.

154

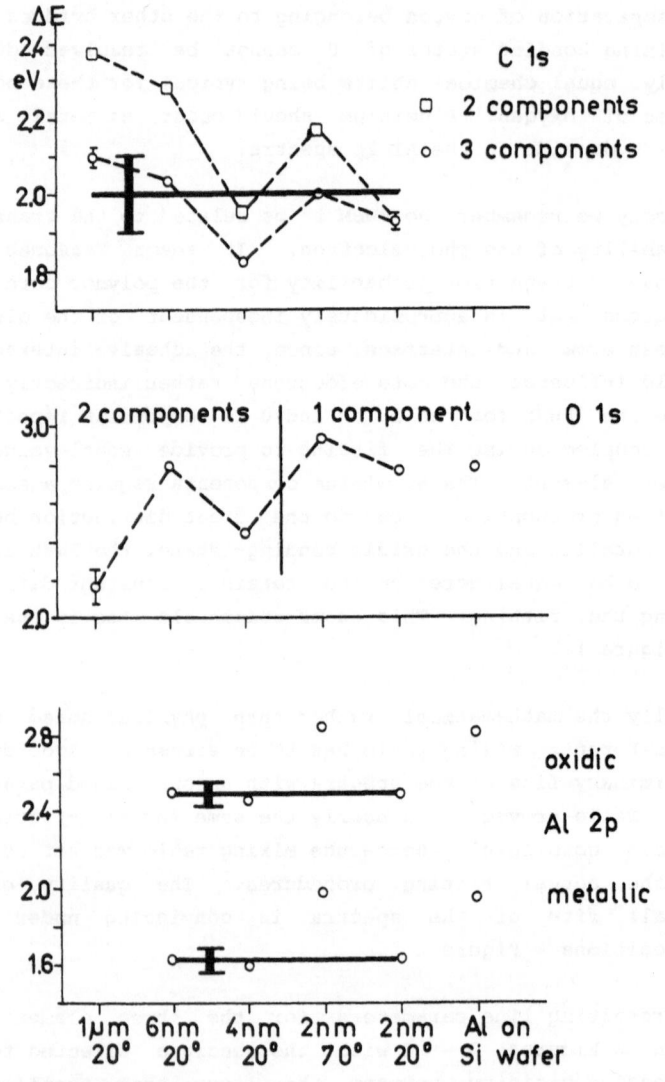

Fig 2: FWHM Δ E of the spectral components obtained by
fitting. For C 1s, the data for a fit with only 2
components are shown for comparison. Full lines
(———) indicate the mean value with the variance (I)
over the spectra.

The behaviour of the FWHM ΔE in Figure 2 tests the validity of the corresponding presumptions. Obviously, the ΔE remain constant for both the oxide and the metallic Al throughout the samples.

This expected behaviour is indeed less clearly revealed by the C 1s components. But comparing the data obtained from the fit of three components with those for two components favours the first variant. Note, however, that both ways yield rather similar ΔE for polymer films thinner than 4nm. For the 0 1s spectra the presumed constant ΔE is difficult to affirm in Figure 2. Looking at the fits not depicted here reveals however that the calculations fail in decomposing two separated components for the polymer layers beneath 4nm thickness.

As a result, besides some evidence for the validity of our approximation concerning ΔE, the first hints of remarkable changes inside the macromolecules near the interface are found in Figure 2.

Two remarks should be put in for the sake of clarity. In Figure 2, as well as in the following ones, the data for the take-off angle of 70^0 do not always fit into the whole picture. This might be due to the different sampling depth. Further, all the Figures contain the substrate data for refrence.

In Figure 3, both the relative inensities of the carbonyl C 1s component and the carbon in the methyl group attached to the ether oxygen in the polymer's side chain drop clearly to the values of the substrate.

Again ca 4nm thick interphase should be deduced at the first glance if there was no carbon contamination from the pump oil. The data for the 0 1s components remove this uncertainty. The relatively intensity of the ether oxygen diminishes continuously with polymer thickness. At 4nm the

156

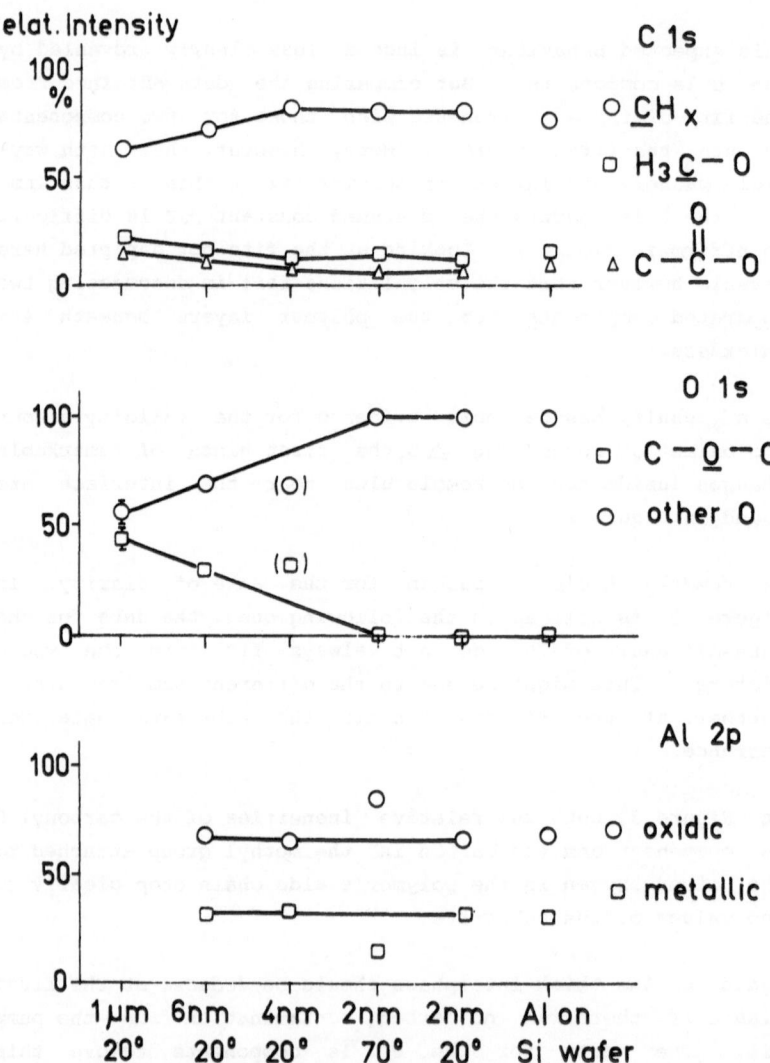

Fig 3: Relative intensity (component area) as a function of
        polymer film thickness and take-off angle.

fitting indicates the end of its resolution capability in the graphs. Therefore the corresponding data points are given in parentheses in Figure 3. Definitely no ether component is resolved from the spectrum for the 2nm thick polymer. For the carbonyl oxygen a similar change can not be excluded but it is impossible to resolve it from the oxidic and hydroxide specied in the interphase. Note that the intensity ratio of the aluminium species is not changed by the presence of PMMA.

All these facts support the existence of an interphase layer with an atomic composition which deviates from that inside the PMMA. These compositional changes are not necessarily caused by adhesive interactions. Figure 4 depicts the data with respect to the binding energy ΔBE determined for the aliphatic C is component (CH_x). This kind of presentation is chosen in order to avoid the well known problem of binding energy referencing. In fact, the binding energy of the C 1s electrons in the carbon species of the polymer side chain enlarges n the boundary layer to the values found for the carbon contamination on the substrate.

A reciprocal and more intense effect is obtained for the oxygen in the ether bridge. For about 4nm, the 0 1s electrons are as loosely bonded to the atom as is the case for the remaining oxygen atoms. A similar behaviour is indicated for the "other 0" which are probably dominated by the carbonyl oxygen in the region above 4nm. Then this trend is covered by the increasing amount of oxygen belonging to the substrate.

For the Al 2p electrons, the spectra reveal no change of the binding energy at all.

4. INTERPRETATION AND CONCLUSIONS

All the trends found for the line shape parameters of the core level photoelectrons support the existence of an

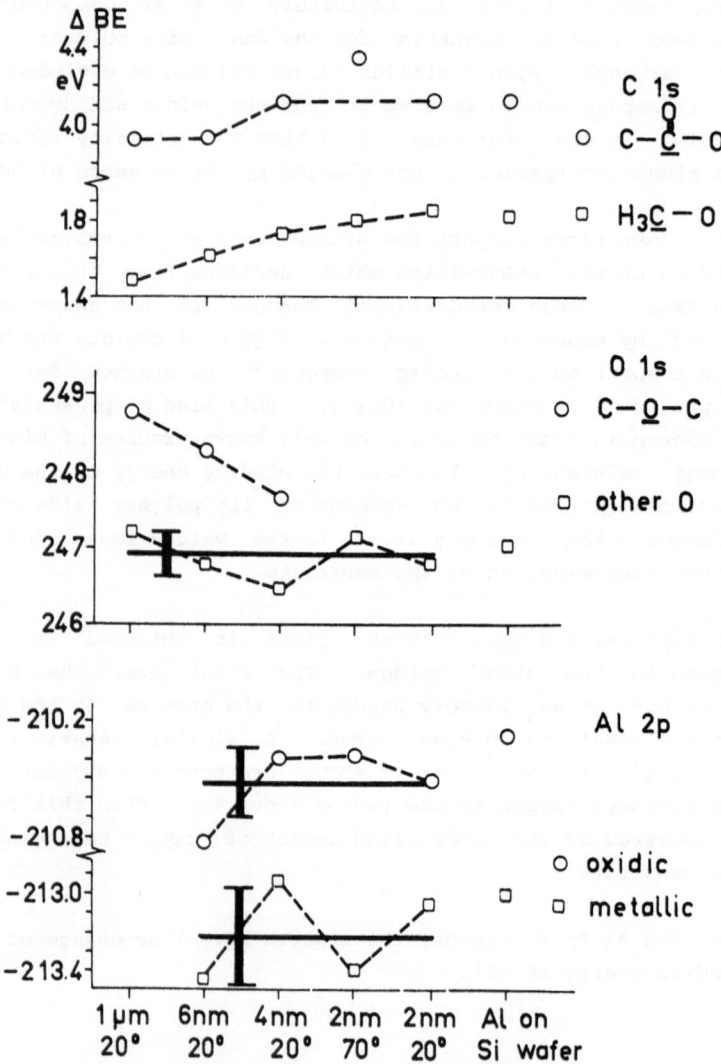

Fig 4: Shift Δ BE = BE − BE(CH$_x$) of the binding energy BE of the spectral components with respect to the binding energy BE(CH$_x$) of the aliphatic (CH$_x$) C is component.———— = mean value; I = variance.

influence of the substrate on the bonding state of both the carbon and the oxygen species in the polymer's side chains. Since no alterations appear in the Al 2p spectra, it seems likely that the substrate should interact with the polymer via the oxygen in the aluminium oxide/hydroxide. The declining binding energy of the O 1s electrons in the interphase meets the semiquantitative picture of extra-atomic relaxation as it is common in the XPS theory. This relaxation enlarges when the bond changes from a covalent into a more ionic one. At this point we remember some well known facts affecting the interpretation of XPS results [7]. The valence electron charge distribution is indeed rearranged by changes of the chemical state of an element but both the shape and the photoelectron cross sections of its core electron orbitals remain essentially unchanged. The binding energies of the core levels, however, are influenced by a rearranged valence charge distribution due to the resulting alterations of the electrostatic potential of the valence electrons. This is the so-called initial state effect contribution to the shift of the core orbital energy (chemical shift). A second important contribution to the chemical shift stems from relaxation processes occuring in the electron distribution at the neighbour atoms in response on the creation of the core hole during the photo process (final state effect).

No adequate explanation exists for the rising binding energy 1s electons in the carbonyl carbon. Obviously, the theoretical concepts available confine the discusion of thre XPS data to qualitative conclusions about interactions revealed. It is even difficult to decide whether the substrate - polymer interaction is an adhesive or weakening one. Such questions afford an improved quantum chemical model of the involved parts of the chain including their interactions with the substrate. In the light of the experimental results such a model should obey the following features. First, the core levels have to be taken into account with respect to the existing interpreation limits for

XPS spectra. Then, the measured changes in core level properties have to be correlated with the valance level state and further with a suitable description of the substrate influence. From our results, this influence may act over the comparably large distance of a few nanometers.

Certainly such a theoretical desription is a task for future research. But some improvement of the spectra handling seems desirable too. The Gauss-Lorentz product function used here represents a mathematical construction rather than a line shape which is definitely related to the features of the photo effect. The introduction of such a suitable line shape function should be accompanied by an adequate removal of the inelastic spectral background. The work on these items is under way. It is coupled to the replacement of the product functions by deconvolution procedures which are the inverse of the processes that happen during the measurement (effect of the spectral function).

The limited physical meaning of the line shape adopted here deserves further comment with respect to the Al 2p spectra. It is well known from the literature that metal XPS peaks are far from being symmetric. On the other hand, the Shirley background tends to symmetrize the spectrum. Only the fact that nearly the same background parameters came out for all the Al 2p spectra encouraged us to resolve them into Gauss-Lorentz components and to discuss the trends of the parameters. Generally, the physically inadequate line shape hinders any quantitative use of the parameters.

Notwithstanding the numerous objections and the complicated unsolved questions, the results represented do not only touch a very interesting area of physics. Moreover, the discussion justifies expectations that XPS in context with an improved theory will be able to provide quantatitive insight into the "chemical" adhesion between polymres and metals.

Acknowledgement

The authors are indebted to Dr D Fanter (IPOC) and G Walther (ZIPC) who created the software. Thanks are also due to Dr A E Hughes (CSIRO) for supplying his program from which many features including the Marquard alogorithm and the line shape function were adopted.

REFERENCES

1. Gardella, J. A., Appl Surf Sci 31 (1988), 72

2. Roberts, R. F., Allara, O. L., Pryde, C. A., Buchanan, D. N. E., Hobbins, N. D. Surf Interf Anal 2 (1980), 5

3. Seah, M. P., in D. Briggs, M. P. Seah (eds) "Practical Surface Analysis by Auger and X-Ray Photoelectron Spectroscopy", Wiley & Sons, New York, 1983, Chapt 5

4. Hughes, A. E., Sexton, B. A., J Electron Spectr & Related Phenom, 46 (1988) 31

5. Shirley, D. A. Phys Rev B 5 (1972) 4709

6. Brandrup, J., Immergut, E. H. (eds) "Polymer Handbook", 2nd ed., Wiley & Sons, New York, Chichester, Brisbane, Toronto, 1975, p IV-38

7. Siegbahn, H., Karlsson, L., Photoelectron Spectroscopy in: W. Mehlhorn (ed) "Handbuch der Physik" vol XXXI, Springer, Berlin 1982 chapt IV

11

INFLUENCE OF ENVIRONMENTAL CONDITIONS
ON ADHESIVE JOINT FAILURE

VERA KOVACEVIC
Institute of Chemical Engineering, Faculty of Technology,
University of Zagreb, Yugoslavia

I MUDRI
Research Institute, Rubber and Footwear Industries, Borovo,
Yugoslavia

I SMIT
"Ruder Boskovic" Institute, Zagreb, Yugoslavia

D HACE, M BRAVAR AND J AZMAN
"Peko" Footwear Industries, Trizic, Yugoslavia

1. INTRODUCTION

Neoprene contact cements have important uses in the shoe industry, particularly for attaching soles and laminating plastics to wood and other surfaces in the automobile and construction industries [1].

Polychloroprene based adhesives were able to establish themselves in many branches of industry because of their relatively high initial strength and very short press times, two factors which considerably shorten production cycles [2].

162

They are known for their capability of bonding together different materials, in short contact times; which is associated with the crystallisation process [3-6].

Polychloroprene is available in a variety of microstructures; the higher the trans-1,4-content the more rapidly the polymer crystalises on deposition from solvent.

The rate of crystallisation markedly affects the cohesive strength and the open time (or time permissible before the two coated surfaces must be brought into contact). Obviously the more rapid and complete the crystallisation, the shorter the open time and the greater the ultimate peel strength of the bond [1].

Highly reactive, tertiary allyic chlorine is present in neoprene as a result of 1-2 polymerization. This comprises about 1.6% of the total chlorine in neoprene and makes the rapid curing of neoprene possible [7].

Formulation characteristics of polychloroprene adhesives, primarily their resin but also other ingredients, should be adjusted according to the desired adhesive function. The request for special substances is therefore inevitable. Quite frequently ZnO and MgO in the role of vulcanising agents[8] are present [7-9].

Metal oxides are essential in vulcanising curing systems. Many metal oxides have a vulcanising affect on chloroprene, but the best system is a combination of magnesium and zinc oxides [7-9]. The linear phenolic resin first reacts with the MgO in the presence of water. Presumably, the magnesium is co-ordinately and collectively bound to the hydroxyls of the methylol and phenolic groups, resulting in a three-dimensional system [1].

Also cross-linking substances are added as well as anti-ageing agents and fillers [4,10]. Alkylphenolic and arylphenolic resins should be regarded as basic ingredients too. These resins with metallic oxides produce adducts of complex nature and assure better thermal stability [11,12].

Many mineral fillers are used in various polymer systems to enhance physical properties in chloroprene compositions [2]. Suitable fillers include magnesium oxide, zinc oxide, magnesium carbonate, silicates and colloidal silicic acid, which improve cohesion.

In this paper the influence of the addition of selected terpene resins on bonding enhancement of the polychloroprene adhesive [10] is investigated, together with their additional dual role as oxidation-inhibiting and stabilizing ingredients [13].

Polyisocyanates as cross-linking substances for all the ingredients which constitute the adhesive, and in interactions between adherend materials and the adhesive, deserve special attention in the context of adhesion phenomena [14].

The joint ageing process occurs through combined deleterious action of air oxygen, moisture, heat and light; especially when cyclic mechanical stress is present.

The great majority of bonded structures are exposed to moist air and if the relative humidity is high, then over a period the strength of joints usually declines. The effect of water on composites has been the subject of many papers in the literature some of which are referred to because of many durability features shared by adhesive joints and composites [15].

Air oxygen significantly attacks some of the participating substances and ingredients in the adhesive. Two basic processes concurrently develop: ie. the degradation of all materials involved together with their structurization [16].

Oxygen, present in the interfacially connected adherend-adhesive-adherend layers, may exert its destructive influence on the durability of the joint in situations where detrimental factors are co-operating. The oxidation kinetics of reacting substances are controlled, not only by the chemical nature of polymeric materials, but also by their morphological peculiarities [17].

A typical polychloroprene vulcanizate eventually fails by ageing processes after exhibiting decrease in elongation and tensile properties. The general nature of a polymer backbone determines its thermal stability.

Adhesive joint failure must involve failure of the adesive in the interfacial area or cohesive failure within one of the joint materials, or some combination thereof [15]. When interpretating the failure of joints, the morpohology of the adhesive polymeric materials under the conditions leading to joint failure needs to be considered. [15].

In this context, the generation and the propagation of cracks within the materials, both adhesive and adherend, (especially at their interfaces), results in enhanced oxygen diffusion into the joint. These cracks, formed in footwear manufacture and/or in their use will in the latter case propagate and quite considerably reduce the time to joint failure in critically stressed footwear.

Fundamental questions such as the mechanism of adhesive bond durability and the locus of the failure can be studied using surface analysis.

DSC analyses and SEM micrographs of the failure surface were used in an attempt to find some correlation between the mechanical properties and the microstructure of the adhesives.

The aim of the experiments presented in this paper is an objective assessment of the facts and situations which are created in strict environmentally-dependent footwear use situations. Single and cummulative-with-time influences of moisture and heat, as well as of the ultraviolet irradiation of the polychloroprene adhesive interfacial film were followed by determinations of the film characteristics.

It is shown here that, on ageing, the adhesive layer is structurally and morphologically altered to such an extent that there is its weakness, not that of adherends, which is responsible for joint failure.

In this paper, the structural changes of the adhesive layers made of two series of adhesives, one series containing the resin (CR_r) in the formulations, the other not (CR) were identified and correlated with differences generated in surface morphology. The assumed joint bonding deterioration mechanisms are discussed, when occuring in the cases of definite environmental situations and when assessed through particular and reliable physical and chemical determinations.

2. EXPERIMENTAL

Materials

The samples for measurements of mechanical properties of mechanical joints, including joint failure determination, were bonded composites as produced in the footwear industry. The adherend materials were natural leather and styrene-butadiene rubber sheets (SBR), in some cases extended with the addition of naphtene-base oil (SBR_{oil}). Adhesives were prepared from neat polychloroprene polymer (CR) and this polymer blended with terpene-phenolic resin (CR_r). Chrome-tanned cattle box-leather for upper parts of shoes containing 4.63% of free fatty substances were used.

Table 1 presents detailed specifications of polymer adherend sheets CR and CR $_r$.

Table 1
Specification of Adhesive Formulations

Adhesive Denomination	Ingredients	%
CR	Polychloroprene	87.5
	Mg	4.0
	Zn	4.0
	Si	4.2
	Desmodur RF	5.0
CR $_r$	Polychloroprene	77.8
	Terpenephenolic resin	27.8
	Mg	4.0
	Zn	3.6
	Si	4.5

To both CR and to CR $_r$ adhesives, 5% by weight of cross-linking agent Desmadur RF was added before application

CR $_p$	Polychloroprene	100
R	Terpenephenolic resin	100

Desmodur RF, used as cross-linking agent is thiophosphoric acid-tris-(para-isocyanato-phenylester).

Silicon dioxide SiO_2 is Vulkasil C, Bayer anhydrous silica having 200 m g of active surface.

Terpenephenolic resin Alresen PK 500 has viscosity of 2000 to 3000 mPa's.

The blending all ingredients with polychloroprene polymer was accomplished by solvent procedure using toluene and petrol fuel 60/80 mixture, for CR samples, with the addition of ethyl acetate for CR$_r$ samples. The solutions were cast on to siliconized glass plates when preparing for experiments on adhesive film.

As is known the polychloroprene variety supplied commercially under the name Baypren C 330 is generally used for adhesive manufacture. It consists of 93.9% trans-1, 4-polychloroprene and of 4.2% trans-isomer. The remainder is 1.3% of 1,2-isomer and 0.7% of 3,4-isomer. M is 2.3 x 10^4.

Investigation Methods

A weatherometer, ATLAS DMC-R (ASTM A-750-68) was used for ageing samples of adhesively bonded composites and adhesive films, the former being exposed up to 60 days and the latter up to 100 hours respectively.

The following ageing conditions were applied:

1. 20 °C, γ =100%
2. 45 °C, γ =50%
3. 45 °C, γ=100% ultraviolet (UV) irradiation

Lap-shear strength and peel strength of the samples which were prepared by the usual standard methods were determined, before and after ageing, according to DIN 53273 standard.

A Zwick 1445 tensile testing machine was used. All tensile testing determinations were made at 20° C and =65% and at crosshead speed of 0.1mm.min using software systems for machine control. The samples were cut into standard specimens conforming to ASTM D-882.

Spectroscopy measurements of adhesive film samples were made
on a Perkin Elmer 457.

A differential scanning calorimeter DSC, Perkin Elmer 7 was
used for observing thermal of changes with heating rate 20
degrees per minute .

TG measurements were made on a Perkin Elmer TGS-2 apparatus
in nitrogen and with a heating rate of 10 degrees per minute
.

Wide-angle x-ray diffraction experiments (WAXD) were
performed using a Philips diffractometer with CuK radiation
in the range $2\theta = 4\text{-}60^{\circ}$, or because the diffraction
intensity of terpenephenolic resin at $2\theta = 54^{\circ}$, becomes zero
(background level), $2\theta = 4\text{-}54^{\circ}$. The component contents were
calculated by the mixing method

An electron microscope, Stereo Cambridge 600, was used for
photomicrography recording of the adhesive film surfaces.

3. RESULTS AND DISCUSSION

Mechanical Investigations of Adhesive Joint

The mechanical properties of adhesive joints before and after
degradation are presented in Figures 1 and 2. The adhesively
bonded composites consisted of leather and elastomer
adherends using polychloroprene adhesives.

The changes which occur up to adhesive joint failure were
followed on samples when subjected to various environmental
degradations that are close to those of ageing in the
exploitation of the materials.

Adherends of SBR elastomer which had previously been oil
extended were included for some cases, as it is usual in

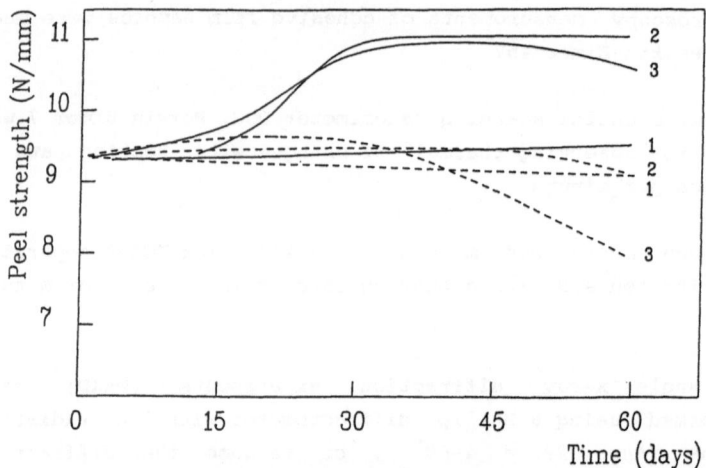

Fig 1. Peel strength of polychloroprene without resin (CR)
adhesive composites Leather-CR-SBR. SBR is without
oil (—) and with oil (--). The ageing dependence
refers to conditions 1-3.

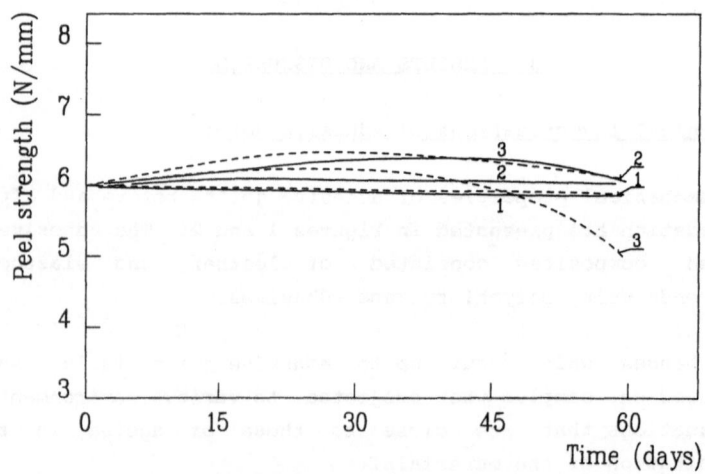

Fig 2. Peel strength of polychloroprene with resin (CR)
adhesive compiosites Leather-CR-SBR. SBR is without
oil (—) and with oil (--). The ageing dependence
refers to conditions 1-3.

footwear industry. By doing so, the detrimental influence of oil to both strength deteminations was noticed. The influence of moisture in degradation conditions 1 was not discovered.

However, by raising the temperature (degradation conditions 2) and especially in the absence of oil a quite significant rise in the mechanical joint strengths was noticed.

This is presented in Fig 1. It can be explained by assumed additional cross-linking in the system. Initial lower strength was found on samples having terpenephenolic resin (Fig 2) but better strength retention after ultraviolet irradiation.

It should be pointed out that adhesive-bonded composites reduce more in strength with ultraviolet irradation (degradation conditions 3) if the oil is present, but only at prolonged exposure of 30 days (Fig 1-2).

The most stable sample proved to be the composite leather-CR_r -SBR $_{no\ oil}$. In this case even at prolonged ultravi olet irradiation, no adhesive joint failure occured (Fig 3).

On the contrary, the least stable composite is Leather-CR -SBR $_{oil}$. It showed its deficiency in containing no terpenephenolic resin and therefore lacking its stabilizing effect. Even worse, this sample includes oil in the elastomer and since oil promotes degradation, it is not surprising that after prolonged exposure of 45 days of ultraviolet radiation, the sample deteriorated seriously.

In the bonded composites, the onset of the degradative process at some exposure time inevitably leads to reduced strength and adhesive joint failure. Samples were flexed up to 3.5×10^4 cycles in the flexometer at several temperatures, and the results obtained are shown in Table 2.

TABLE 2. MECHANICAL CHARACTERISTICS OF ADHESIVE JOINTS AFTER 3.5 · 10⁴ FLEX. CYCLINGS ON VARIOUS TEMPERATURES WITH POLYCHLOROPRENE ADHESIVES (CR) AND (CR$_r$)

ADHESIVE COMPOSITES	LEATHER-CR-SBR					LEATHER-CR$_r$-SBR				
	BEFORE	AFTER				BEFORE	AFTER			
	TEMP(°C)	-20	0	+20	+40	TEMP(°C)	-20	0	+20	+40
G-Shear (N/mm²)	1.32	1.43	1.36	1.38	1.32	1.10	1.25	1.19	1.23	1.22
F-max (N)	528.80	572.45	543.55	550.75	526.19	440.38	500.06	477.15	493.12	487.74
Work to break (Nm)	70.47	75.09	74.56	72.35	71.75	55.16	61.04	57.56	58.61	60.65

TABLE 3

Mechanical characteristics of adhesive composition films, (CR) and (CR$_r$) before -0- and after -100- UV degradation.

Mechanical Characteristics	ADHESIVE FILMS			
	CR		CR$_r$	
Degradation time (hrs)	0	100	0	100
Young's Modulus (MPa)	103.4	21.6	80.5	24.6
σ (N/mm)	23.2	7.7	19.4	11.5
ε max (%)	599.2	313.8	736.6	564.2
σ fracture (N/mm²)	23.2	7.6	19.4	11.4
ε fracture (%)	576.7	278.4	718.8	517.7
σ yield (N/mm²)	4.2	2.5	3.6	1.8
ε yield (%)	14.4	23.0	12.6	21.1
Work to break (Nm)	54.0	9.0	60.6	29.7

σ – stress   ε – strain

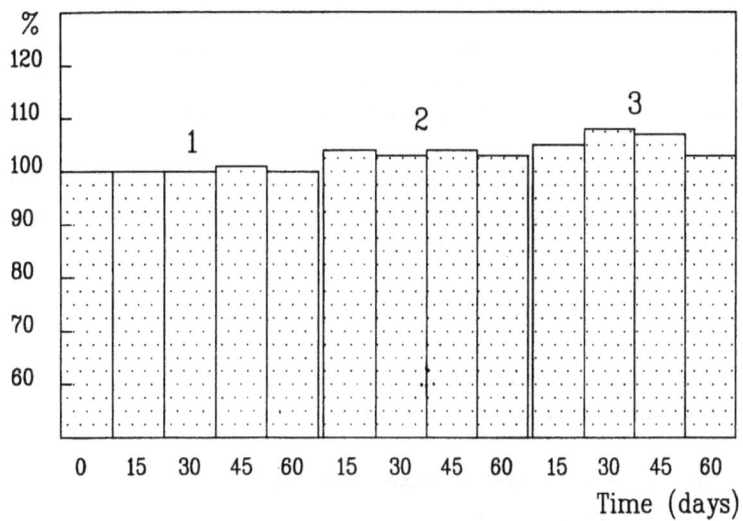

Fig 3. Relative changes in peel strength at various ageing
conditions 1-3, for the most stable adhesive composite
Leather-CR WITH RESIN -SBR WITHOUT OIL

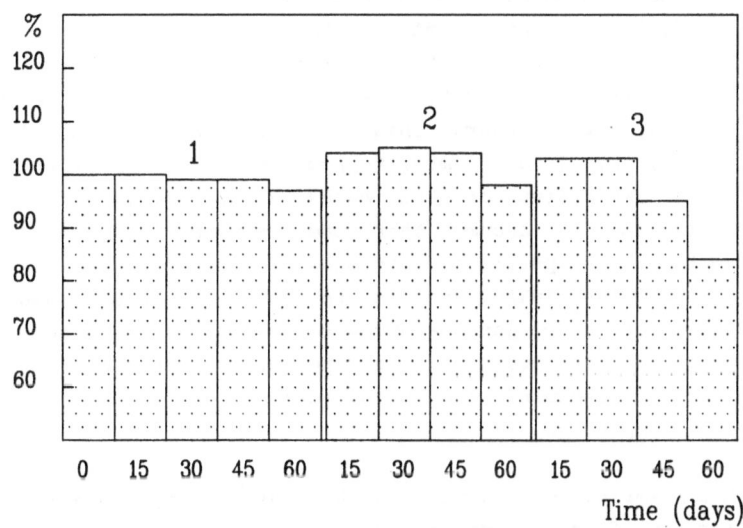

Fig 4. Relative changes in peel strengths of various ageing
conditions 1-3 for the most unstable adhesive compsite
Leather-CR WITHOUT RESIN -SBR WITH OIL .

Obviously it is not possible to detect the influence of this kind of mechanical stress with reduced numbers of flexing. However these values are found to be raised, to a small degree, with aged samples, especially at lower temperatures.

The measurements at low temperatures revealed the reliability of using polychloroprene adhesive for these candidates as well as at higher temperatures.

Investigations of Adhesive Films' Mechanical Properties

Adhesive films' mechanical strengths, both for (CR) and (CR_r) compositions are presented in Table 3.

As usual with elastomers, here also stress-strain relations are nonlinear [20].

Higher values of tensile strength and of flexural moduli are the result of the co-existence of crystalline and cross-linked structures as in the case of the sample (CR). The (CR_r) sample has a minor share of the crystalline structure, as detected by the crystallographic measurements, but the elasticity is better, as well as the work to break. Since in all other regards this sample proved to be superior in ageing, this finding may, to some extent, be worthwhile for the assessment of the adhesive stability. It is generally true to assume that the modulus is lower in degradation, but parallel processes of the structurization and of the cross-linking are also to occur. The predominance of one, or a proper balance between these processes, affecting the structure of the polymeric material would inevitably determine the mechanical properties of the adhesive compostion.

Drastic lowering of modulus with degradation, relatively smaller for film with resin (CR_r) leads to the conclusion that photo-oxidative degradation took place (Table 3).

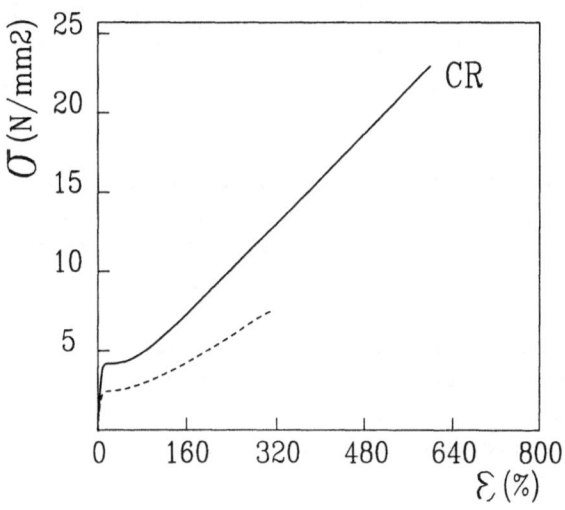

Fig 5. Comparative mechanical curves of polychloroprene
adhesives without resin (CR) before (—) and after
ultraviolet degradation (--).

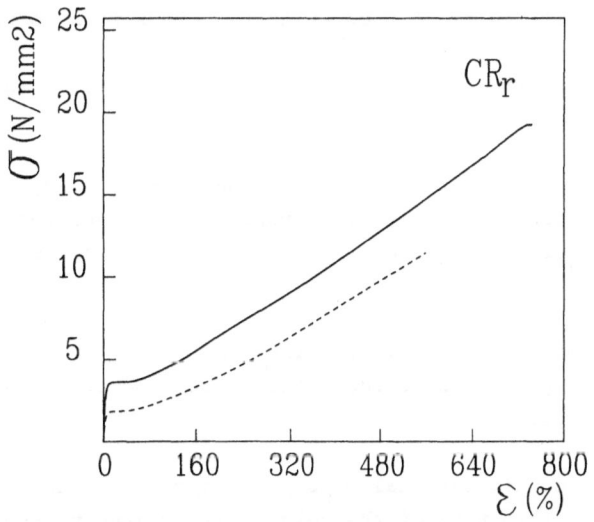

Fig 6. Stress-strain curves of unaged (—) and aged (--)
polychloroprene adhesive film with resin (CR).

The adhesive systems with resin shows lower inital strength characteristics but more stability because on ageing they have been less degraded than the adhesive systems without resin.

The stress σ relating to a given strain ε , increased with increasing amounts of cross-linked bonds as in sample CR. The modulus, E, also increased with increase of cross-linked bonds.

The contribution of hydrogen bonding and other forces to the modulus was considerable. As the temperature increases these secondary bonds disappear, causing the drop in E values [20]. Degradation processes lower the modulus and other mechanical characteristics.

Spectroscopic Investigations of the Films

Spectroscopic analysis of the changes in the chemical and physical structure of the adhesive was made on the basis of the correlations of the cummulative absorption bonds with these given in the literature [8,21-26] for the polychloroprene polymer, terpenephenolic resin and other ingredients.

The main bands at 3500, 2920, 1660, 1580, 1430, 1100, 820 cm $^{-1}$ assigned to polychloroprene were identified, as was the band for SiO_2 (1100cm $^{-1}$) common to the samples in question.

In addition the bands (1740, 1450 and 1380 cm $^{-1}$) in the adhesive spectra where identified where the terpene-phenolic resin was present.

Trans-1,4-configurational polychloroprene bands [8] were noticed as this type of the polymer is present to the exttent of 93.8%. The other structures were cis (4.2%) and 1,2 (1.3%) and 3,4 (0.7%) do not interfere with this.

Some other bands too, namely the one at 952 and the other at 782 cm^{-1} [8] should be taken into consideration because they are associated with the degree of crystallinity of the polychloroprene which is to be of practical importance as it is related to the contact adhesivity of polychloroprene adhesives.

The absorption band 925 to 934 cm^{-1} related to the existence of 1,2 olefinic groups, neighbours to allychloride group [21], may also play some minor role, since this band appears to reactive adhesives [8]. Identification of structural changes, as the consequence of ageing processes, could lead to an understanding of the mechanisms of degradation and explain the physical and chemical properties of adhesive films.

Only in the severe conditions of ageing, conditions (3), was the influence of moisture, heat and ultraviolet irradiation confirmed by IR spectra, whereas in other cases this was not so.

New bands did appear as well as the broadening and shifting of the existing bands at 1550-1610-1740 and at 1090cm^{-1} was noticed, all of which is possibly due to ultraviolet irradiation. These, evidently, cumlative maxima resulted from oxidation and degradation processes. The disappearance of formerly existing bands at 1300 cm^{-1} and of crystalline-part bands at 952 and 782cm^{-1} maybe due to degradation of the polychloroprene. It is known that in crystallisation of this polymer, bands at 952 and 782cm^{-1} assigned to the 1-4trans structure became more intense [8]. In other words, the dimishing of these absorption bands indicates the degradation reducing the crystallinity. The band at 1025cm^{-1} maybe assigned to cyclopropyl structure [27,28] formed at structurization in ageing.

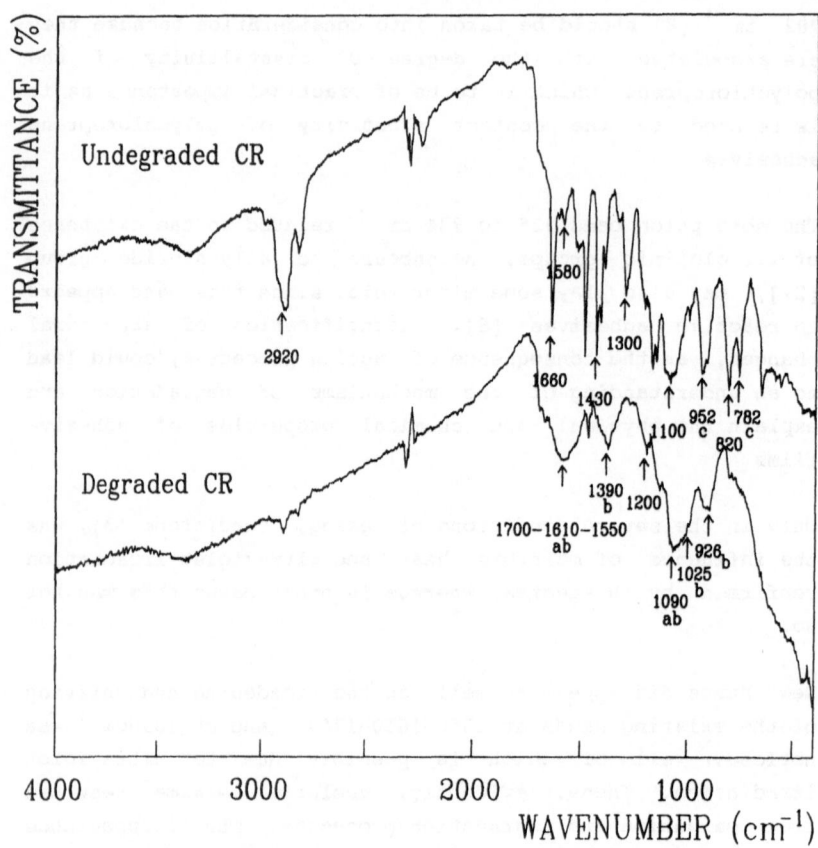

Fig 7. Comparative IR spectra of undegraded polychloroprene
adhesive (CR) and of degraded ones after ultraviolet
irradiation (a-shifting of maximum of absorption;
b-appearance of new maximum; c-crystalline band)

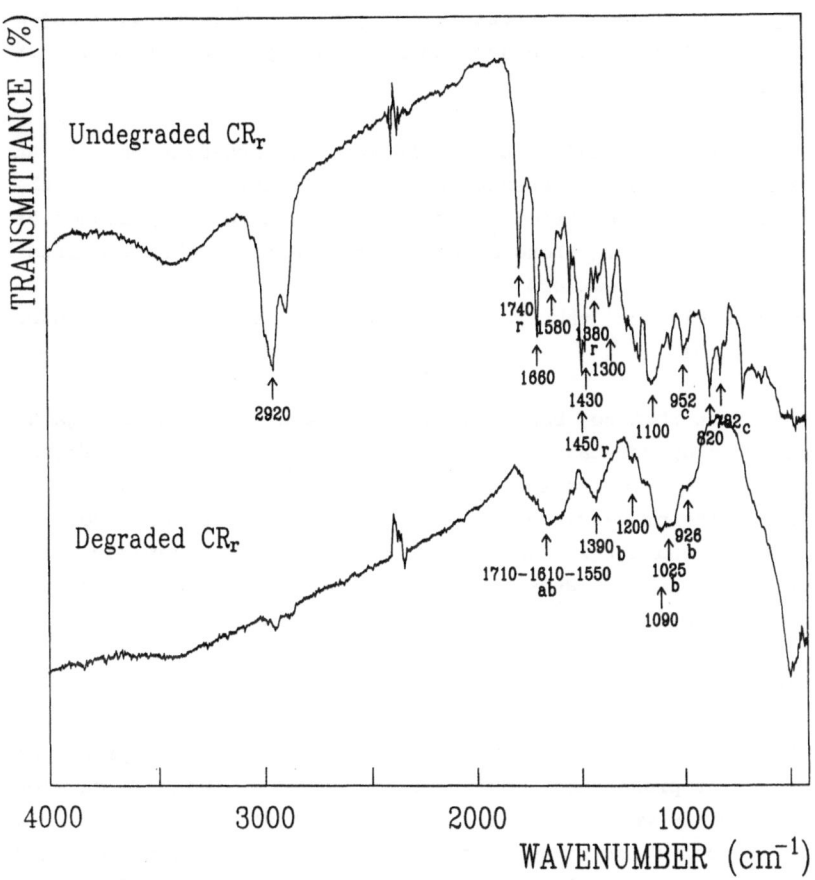

Fig 8. Changes in IR spectra of polychloroprene adhesive composition with resin (CR$_r$).

Structural changes in polychloroprene adhesives containing terpenephenolic resin do not alter significantly (Fig 8). It may be concluded that oxidation path and structural changes in polychloroprene are not influenced by the presence of resin.

It maybe concluded that ultraviolet irradiation of an adhesive film is detectable by the transformations of the existing IR absorbtion bands along with the appearance of some new bands. Photo-oxidative degradation, observed after 30-45 days of weatherometer adhesive film ageing, may be responsible for the loss of strength in adhesive-bonded composites.

The fact that no basic differences were found after ageing polychloroprene-based adhesives which contained terpenephenolic resin in their formulation, indicates that the ageing processes occur in the polychloroprene polymer itself, and not in the resin. The resin may, therefore, act as a resistant substance and stabilizing agent in the adhesive system.

Ageing Mechanisms

The investigations on spectroscopic changes in ageing of polychloroprene-based adhesives allow some ideas to develop to explain what is occuring in an adhesive joint; primarily in its polychloroprene adhesive part, under the combined action of ultraviolet irradiation, heat and water. Oxidation ageing, enhanced by heat and/or ultraviolet irradaition is in question here as the most relevant type of ageing [16,25].

The stability of the polychloroprene adhesives is also influenced by the ingredients, primarily by the added synthetic resin [8] in the adhesive formulation. Two basic aspects of degradation changes appear concurrently, the polymer destruction and its structurization. The adhesive joint strength reduction in the footwear or other similar commodities may be caused by oxidation cleavage of the polychloroprene by radical mechanism [16].

In the inital stage of the oxidtion process the oxygen reacts with the double bond units in the polymer [29].

$$
\sim CH_2 - \underset{\underset{Cl}{\mid}}{C} = CH - CH_2 \sim \quad \xrightarrow{O_2} \quad \sim CH_2 - \underset{\underset{O-O}{\mid}}{\underset{\mid}{C}} - \underset{\mid}{CH} - CH_2 \sim
$$

Peroxide structures are readily cleaved forming free radiacal speciies which initiate both polymer degradation and struxrization processes via crosslinking. During the degradation processes the oxidation products containing carbonyl and/or hydroxyl groups are formed:

$$
\sim CH - \underset{\underset{Cl}{\mid}}{C} = CH \diagup^{CH_2 - CH_2} \diagdown_{C} = CH - CH_2 \sim \quad \xrightarrow{ROO^\bullet}
$$

$$
\sim CH - \underset{\underset{Cl}{\mid}}{C} = CH \diagup^{CH_2 - CH_2} \diagdown_{\underset{\underset{OOR}{\mid}}{C}} = CH - CH_2 \sim \quad \xrightarrow{2O_2}
$$

$$
\sim CH_2 - \underset{\underset{O^\bullet}{\mid}}{C} = CH \diagup^{CH_2 - CH_2}_{\diagdown O ---- O \diagup} \diagdown_{\underset{\underset{OOR}{\mid}}{C}} = CH - CH_2 \sim \quad \longrightarrow
$$

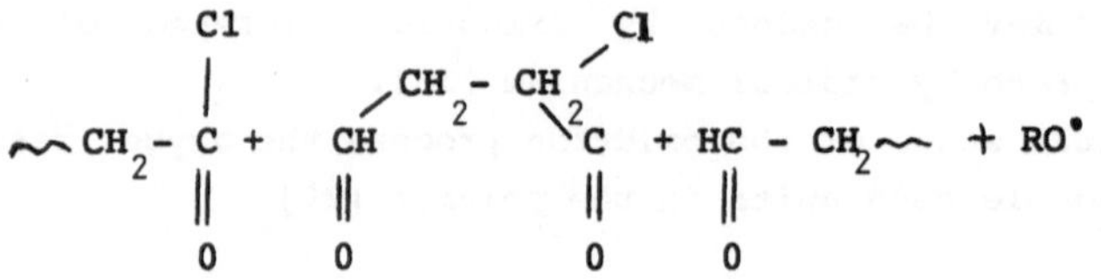

~CH$_2$-C + CH$\diagup$CH$_2$-CH$\diagdown$C + HC-CH$_2$~ + RO$^\bullet$

with Cl on the C (left) and Cl on the CH (right), and ‖O groups below.

Scheme 1

For polychloroprene, the mechanisms of oxidation ageing may be represented as shown in Scheme 1.

The changes in spectra of degraded polychloroprene adhesive films (Figs 7-8) in the 1700cm^{-1} band may be associated with the formation of oxidized products containing carbonyl, aldehyde and keto groups. The 1390cm^{-1} band changes may be assigned to the deformational oscillations of hydroxyl groups. Oxidation kinetics is also dependent upon the polymer supramolecular structurization [17].

As known in the special-purpose polychloroprene, the 1,2-isomer structure is favoured for adhesive preparation [8]. The chlorine in the allylic form is labile and responsible for curing and crosslinking reactions. The structurization of the polychloroprene is also believed to occur under the influence of air humidity, whereby the chlorine atom, in the form of hydrogen chloride is split-off in very small quantities. Ether bridge [30] formation by hydrolysis [16] between chains may be interpreted by examining IR spectra of degraded polychloroprene at 1200cm^{-1}.

Thermal Analysis of Adhesive Films

The glass tranistion temperatures (T$_g$) are shown in Table 4.

Pure polychloroprene sample (CR$_p$) and adhesive composition without resin (CR) have similar curves with relaxation (Fig

9). The added resin lowered the crystallinity and raised the T$_g$. During degradation the T$_g$ is not shifted; it only influences the shape of DSC curves.

Thermo-gravimetric analsyis of samples (Table 4, Fig 10) could be useful for identifying the influence of degradation on the thermal stability of components and polychloroprene adhesive compositions. Tg curves of degraded samples are shifted to lower temperatures as a consequence of polymer scission but at higher temperatures they are moved to some higher residual weight because of crosslinking processes.

In the adhesive composition without resin (CR) the degradation processes overlap the structurization.

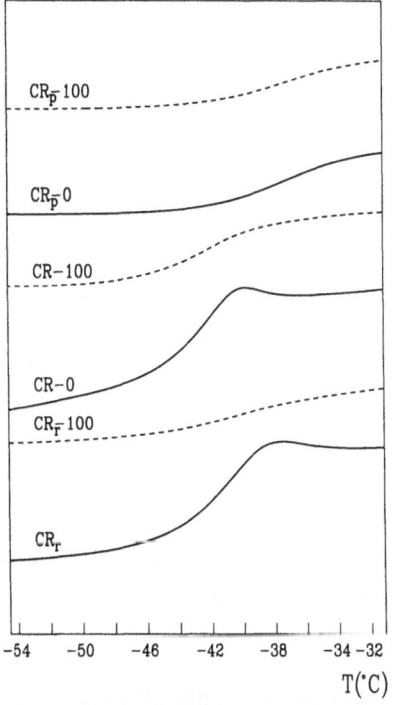

Fig 9. Compariuson between normalized DSC curves of pure polychloroprene component (CR$_p$) and adhesive compositions (CR and CR$_r$) before -0 and after ageing -100.

TABLE 4

DSC and TG results of adhesive component pure polychloroprene (CR) comparing with adhesive compostion (CR) and (CR$_r$) before -0 and after ultraviolet degradation -100

Samples Adhesive Component	DSC T_g (°C)	ΔC_r (J/g-c)	TG WL(%)[1]	RW(%)[2]	T_1(°C)[3]	T_2(°C)
CR$_\rho$ -0	-41.7	0.367	80	20	189	590
CR$_\rho$ -100	-40.9	0.072	77	23	135	586
Composition						
CR -0	-43.0	0.321	61.5	38.5	157	605
CR -100	-41.8	0.212	63	37	105	615
CR$_r$ -0	-38.1	0.224	70	30	114	500
CR$_r$ -100	-37.3	0.150	71	29	115	613

1 weight loss
2 residual eight
3 TMA maximum

X-Ray Diffraction Results

The lower crystallinity of polychloroprene adhesive with resin (CR $_r$), observed by IR spectra and lower initial adhesive joint strength, have to be confirmed by WAXD measurements.

Polychloroprene, 1,4-poly(2-chlorbutadiene), trans, CR as a compound in adhesive as semicrystalline polymer containing 24.2% crystal CR phase according WAXD; with an orthorombic

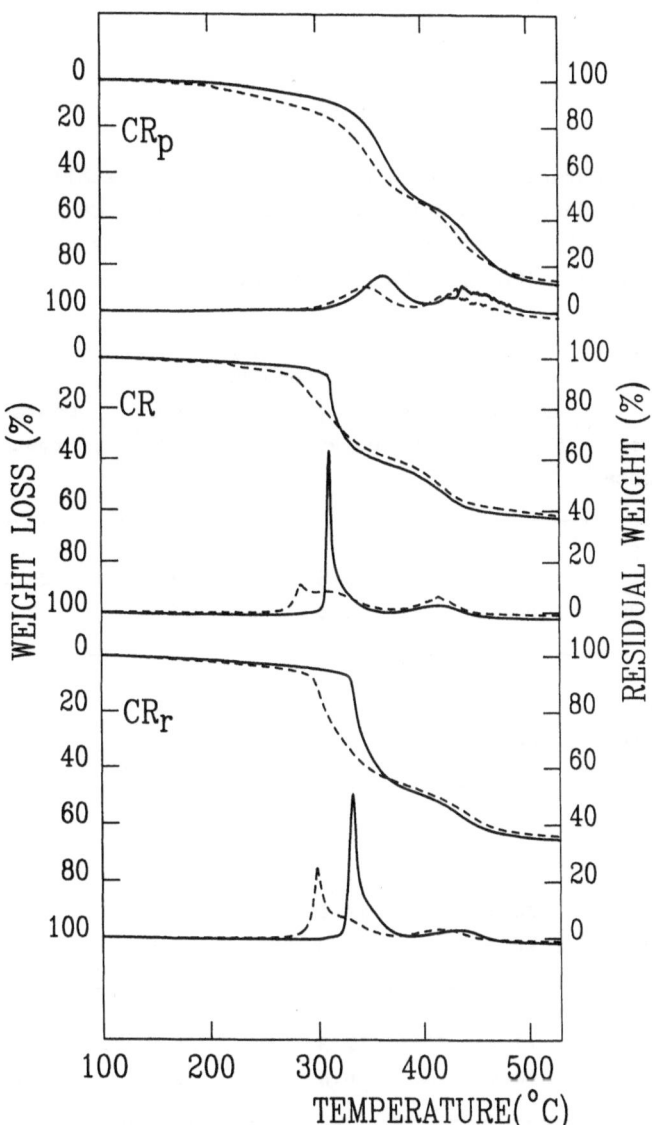

Fig 10. TG curves for polychloroprene component CR and adhesive compositions CR and CR before (⟶) and after ageing (--).

unit cell [31,32] (a=8.84 Å, b=10.24 Å, c=4.79 Å), space group $D2_1 2_1 2_1$ (table 5).

TABLE 5

Relative contents of amorphous phase W_α and crystalline phase W_{cr} of adhesive components: polychloroprene (CR$_p$) and terpenephenolic resin (R) comparing with adhesive composition (CR) and (CR$_r$) before -0 and after ageing -100.

Samples	WAXD Results			
	Amorphous	Crystalline Phase		Degree of
Adhesive Components	W_α	W_{cr}(polychlorop)	W_{cr}(Addit)	Cryst (%)
CR$_p$ -0	30.2	9.6	60.2	24.2
CR$_p$ -100	66.4	0.0	33.6	2.3
R -0	100	-	-	-
R -100	100	-	-	-
Composition				
CR -0	72.9	14.6	12.5	-
CR 100	92.1	0.0	7.9	-
CR$_r$ -0	77.9	12.3	9.9	-
CR$_r$ -100	90.0	0.0	9.1	-

The additives (alumino-silicate, talc, clinochlore (A), ZnO, MgO were added to semicrystalline polychloroprene. The crystalline phase present was: talc with the following values d = 9.50, 4.79, 3.14, 2.85 A and clinochlore values d = 14.48, 7.20, 3.58, 2.89 were identified.

Crystalline phase of polychloroprene is assigned by the main values: d = 4.47, 4.11, 3.36, 3.08 and 2.71 Å. Crystalline phase (CR) was destroyed during the ageing completely in all samples. One part of alumino-silicates additives was amorphized too. Structural changes of amorphous components and polychloroprene adhesive compositon couldn't be resolved from diffractograms because of their complex natures. It is only evidently that the big maximum (d=5.56 Å) of amorphous terpenephenolic resin broaded in the direction of smaller Bragg's (d) spacings during the ageing. (d=5.50 Å)

WAXD curves of pure polychloroprene (Fig 11) show how the crystalline phases are destroyed during the degradation and also, but not so intensive, the crytalline part from alumo-silicate additives.

The influence of degradation processes on morphological structures of adhesive composition films in Fig 12 confirm that the lower content of crystalline phase from polychloroprene component (CR) and additives (A) and adhesive composition with resin (CR$_r$) is the consequence of stabilisation of the system with amorphous terpenphenolic resin. The degradation processes cause the complete disruption of the polychloroprene crystalline phase and the degree of crystallinity falls to zero.

As a consequence, the amorphous phase grows as the result of the processes previously mentioned as well as is partly amorphization of crystalline phase in additives in polychloroprene (Table 6). The content of crystalline phase in the additives is lowered from 12.5% to 7.9%.

188

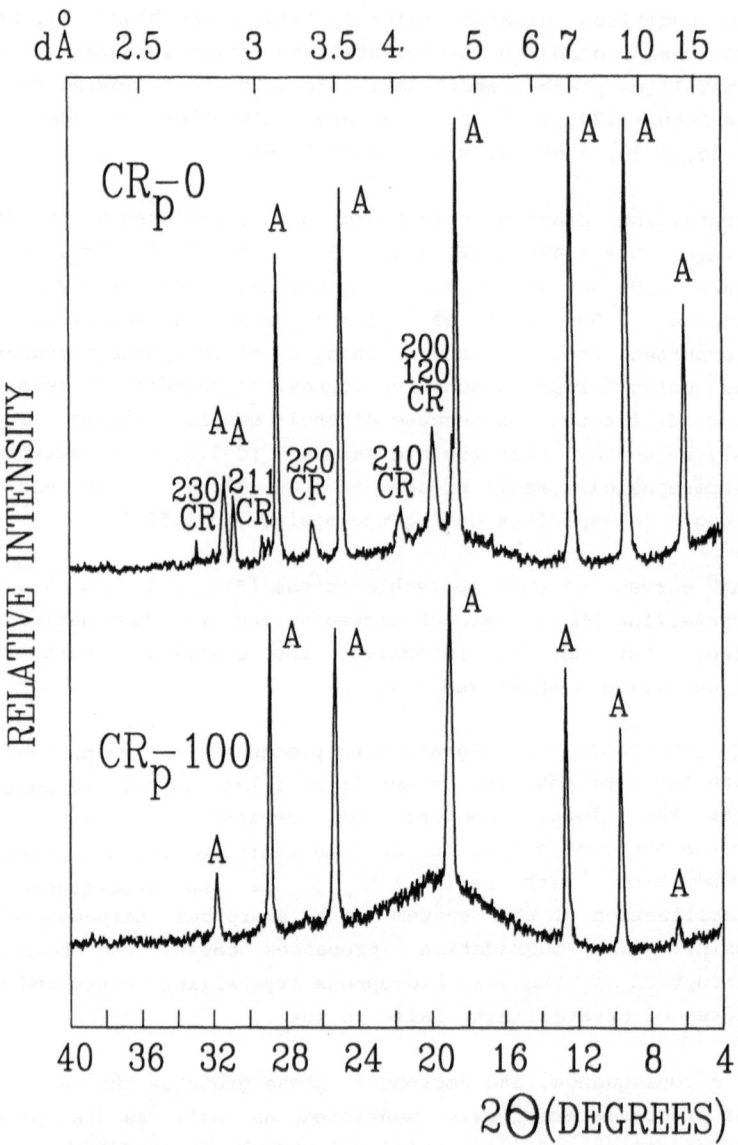

Fig 11. Changes in crystalline peaks (CR) of polychloroprene component in adhesives (CR$_p$) and alumo-silicate additives (A) before -0 and after degradation -100.

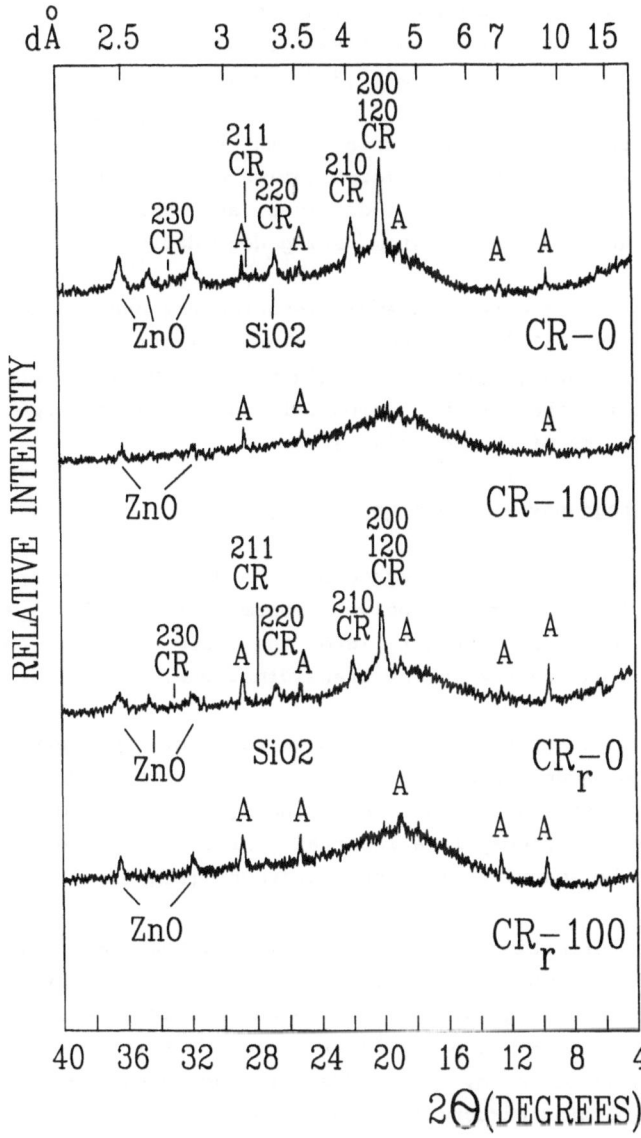

Fig 12. WAWD Curves of polychloroprene adhesive composition
without resin (CR) and with resin (CR$_r$) before -0
and after degradation -100.

In the adhesive composition with resin (CR $_r$) the result is contrary to this, the additive crystalline content remains unchanged during degradation (9.9 - 9.1%; Table 5).

This behaviour could be explained as the influence of resin on the lower degrees of amorphization of crystalline content according alumosilicate additives and in a way some stabilization of this system during degradation.

On the basis of crystallogaphic measurements one can conclude that the crystalline phase of alumosilicate additives in polychloroprene is less amorphized when the terpenphenolic resin is in the adhsive composition and causes cooperative stabilization of the adhesive system.

Microphotography of Adhesive Surfaces

The morphology of the surface of the degraded polymer using samples in the form of polychloroprene-based films should possible give an insight into the role of the surface condition during the ageing transformations.

The addition of the polyisocyanate results, not only in its exclusive crosslinking reaction in the process of bonding together materials, but additioally in its ability to structurize sufaces, in a way that fibrils are produced as revealed by scanning electron microscopy (Fig 13).

Better definition of the fibrilar structure formed is observed in samples produced by adhesives with terpenphenolic resin. This ingredient, as shown earlier, assures the adhesive being more resistant in drastic environment. Terpenphenolic resin, when added to the adhesive, enhances fibrile-forming structurization of the surface (Fig 14).

In degradation especially if ultraviolet irradation is present, the disruption of the fibrilar structure was observed, however to a minor extent if terpenphenolic resin

CR adhesiv

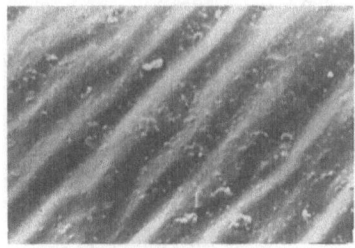

CR_r adhesiv

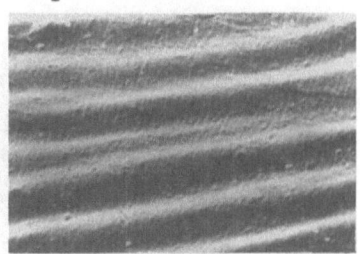

With crosslinking agent

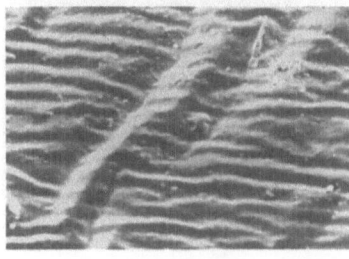

With crosslinking agent

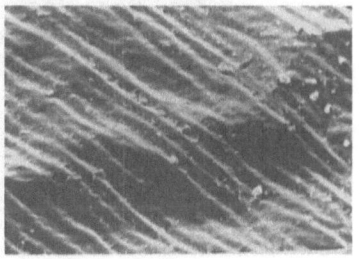

Fig 13. Influence of crosslinking agent (Desmodur RF) on micrographs of polychloroprene adhesive surfaces (CR and CR_r)

ADHESIVE COMPOSITION

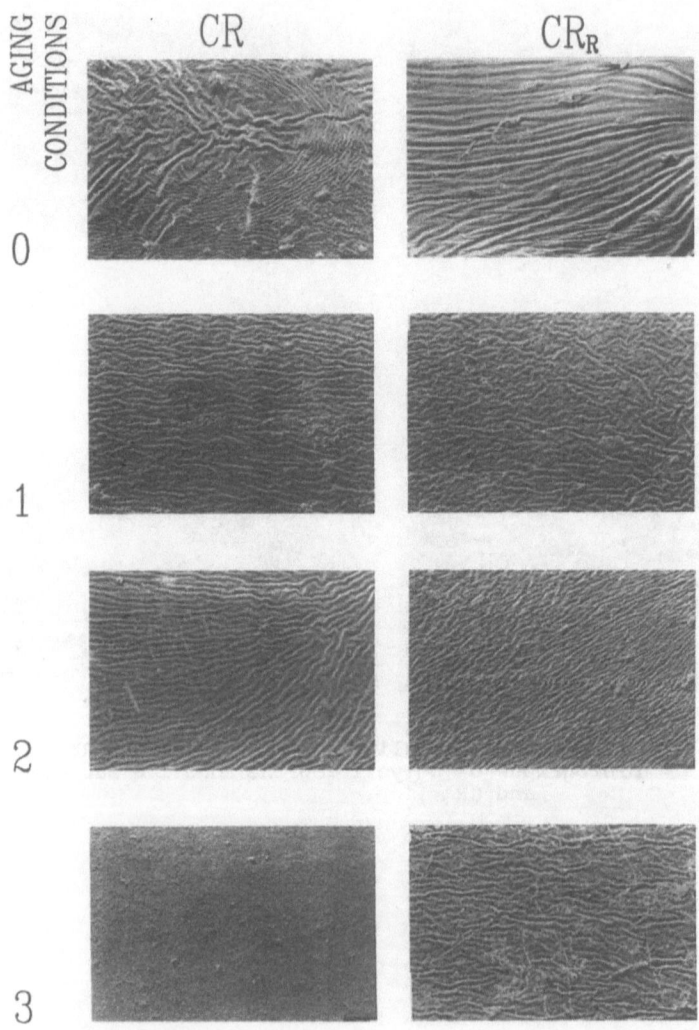

Fig 14. SEM micrographs of adhesive composition films without resin (CR) and with resin (CR$_r$) in various ageing conditions 1-3

was present in the adhesive. The electron microscopy is therefore an indispensible method for the assessment of the adhesive film surface condition before or after degradation.

CONCLUSIONS

The influence of degradation conditions, especially of ultraviolet irradiation cause adhesive joint failure to occur when polychloroprene adhesives with leather and styrene-butadiene elastomer as adherends were used. The characteristics of adherend, as well as of the ingredients in adhesive compsitions, have significant influence on changes of properties in ageing conditions. During the ultraviolet irradiation the polychloroprene component in adhesive composition was degraded and complete dissappearance of its crystallinity phase was observed.

Terprenephenolic resin was found to act as stabilizing agent. Air moisture should be regarded as environmental factor of minor importance as well as temperature, in the ageing processes of polychloroprene-based adhesive joints.

Therefore ultraviolet irradiation is the predominant factor in combined artifical ageing process, where, presumably, auto acceleration effects play important role and where terpenphenolic resin acts as stabilizing agent.

REFERENCES

1. Mark, H.F., Bikales, N.M., Overberger, Ch 6, Menges 6, Encyclopedia of Polymer Science and Enginering, Wiley-Interscience Publication, p556, 1985

2. Lee, M.C.H., J. Appl Polym Sci., 33 (1987) 2473

3. Malers, L.J., Kalenin, M.M., Mehanika polimerov, 3 (1976) 420, 424

4. Landan, M., Adhesion 3 (1980) 64

5. Matulewicz, C.M., Snow Jr., A.M., Adhes Age., 25 (1981) 40

6. Snow Jr., A.M., Adhes Age., 23 (1980) 35

7. Murray, R.M., Thompsdon, D.C., The Neoprenes, E.I. Du Pont de Nemours and Co (Inc), Wilington p65, 1964

8. Hummel, Scholl, Atlas der Kunstoff-Analyse, Band 1, Verlag, C.H., Munchen p151-152, 1968

9. Dogalkin, B.A., Donstov, A.A., Cherochnev, V.A., Khimiya Elastomerov, Izd. Khimiya, Moskva, p 108, 1981

10. Schunance, E., The Effect of Synthetic Resin on the Properties of Polychloroprene Adhesives, ATA KH I/HT, Weil p1-7, 1973

11. Hultzsch, K., Farbe und Lack 77 (1971) 1165

12. Kwiatek, J.F., Adhes Age., (1988) 28

13. Wagner, H.F., Lackkunstharze, Carl Hanser Verlag, Munchen, p3, 1971

14. Levenka, P.I., Kozevnoobuvnaya promislenost 11 (1973) 47

15. Kinloch, A.J., Durability of Structural Adhesives, Applied Science Publishers, New York 1983

16. Piotrovski, K.B., Tarasova, Z.N., Starenie I., stabilizacia sintetiski kaucukov i vulkanizatov, Khimija, Moskva, p13, 103 1980

17. Bogaevskaya, T.A., Vysokomol.Soedin 14 (1972) 1552

18. Chaw, K.W., Geil, P.H., Polymer 25 (1985) 490

19. Zevin, L.S., Zavyalova, L.L., Kolichestvenniy rendgenographicheskiy prazoviy analiz, Nedva, Moska, p37 1974

20. Dzierza, W., J Appl Polym Sci., 22 (1978) 1331

21. Dexant, I., Infrakrasnaya spektroskopiya polimerov, Khimiya Moskva p371, 1976

22. Ferguson, R.C., J Polm Sci., 2 (1964) 4735

23. Mamani, S., Brigodiot, M., Marechal, E., J Appl Polym Sci., 29 (1984) 4081

24. Ferguson, R.C., Analyt Chem., 36 (1964) 2204

25. Miyata, Y., Atsumi, M., J Polm Sci., Part A 26 (1988) 2561

26. Baker, T.E., Fix G.L., Judge, J.S., Adhez Age., 24 (1980) 39

27. Golub, M.A., Pure Appl Chem., <u>30</u> (1972) 105

28. Golub, M.A., Macromolecules <u>2</u> (1969) 550

29. Garmonova, I.V., Sinteticskyi kaucuk, Khimiya, p368-386, 1976

30. Morand, I.L., Rubb Chem Technol., <u>47</u> (1977) 1094

31. Clews, C.J.B., Proc Roy Soc., A180 (1942) 100

32. Bunn, C.W., Proc Roy Soc., A180 (1942) 40

12

DEVELOPMENT AND STUDY OF HYDROPHILLIC EPOXY BASED ADHESIVES.

D A Tod and S J Shaw
RARDE
Waltham Abbey
Essex
EN9 1AX

INTRODUCTION

Military equipment in general is expected to be exposed to a
wide range of environments. These can range from the very cold
areas of central Norway to the very hot dry regions of the Arab
states. An important component when determining the life of
materials is the presence of water which generally acts to
enhance the rate of degradation of equipment. This is certainly
true in the case of adhesively bonded systems. Water has been
shown to act in many ways to degrade the performance of a joint.
It can act directly upon the metal substrate and change its
nature such that the adhesive is forced from the surface. An
example of this would be an aluminium oxide surface which became
hydrated in time. A second method is the simple displacement of
the adhesive by water, as there is usually only a physical bond
between the adhesive and the substrate this displacement process
can proceed in time. A third method is the degradation of the
adhesive composition itself, this may be of a reversible nature
whereby moisture is absorbed the adhesive is plasticised and no
longer has the strength to resist service loads. This
plasticisation may be reversed when the adhesive is dried and the
joint strength may recover. However there are effects such as
microcracking of the bulk of the adhesive that cause irreversible
damage. There are two general approaches to overcome these
problems one is to improve the adhesive substrate interface. An

196

example of this approach would be the use of a primer such as a silane which may form a chemical linkage between the adhesive and the metal. The second approach is to develop moisture resistant adhesives which do not allow the transmission of moisture to the interface region. The ultimate aim would be to develop an adhesive system which had a very low diffusion rate for moisture together with a very low equilibrium moisture level. In practice either of these conditions may be appropriate for a moisture resistant adhesive systems.

One of the most widely used adhesive system are those adhesives based upon epoxy resins which combine good mechanical properties with ease of use. In this paper we describe a series of experiments to develop tough water resistant adhesive based upon epoxy resins. This paper covers the initial stages of the research program in which combinations of resin/curing agents are exa mined to determine the most promising system as a basis for use as an adhesive. The modifications made concern the use of fluorinated resins and curing agents which have a natural resistance to moisture. These system were developed in the United States by the Naval Research Laboratory (1-5).

EXPERIMENTAL

Materials

The fluoroepoxy resins used in this study have the general structure as shown in figure 1. Except for the case in which n=0, all members of this resin series are clear, colourless syrups. As the fluorine groups which provide the hydrophobicity are fairly remote from the reactive end groups these resins react in a similar fashion to diglycidyl ether resins. The major problem with such materials is that they are incompatible with common curing agents such as aliphatic amines. To overcome this problem two types of curing agent have been developed (5), the silicone amines and the fluoroanhydrides. The silicone amines have the general structure as shown in figure 2 and it should be noted that these materials have a similar hydrophobicity to the base resins. A number of fluoroanhydride curing agents have been developed and two of these are shown in figure 3. It is necessary with these curing agents to use an elevated temperature cure

FLUOROEPOXY RESIN/CURING AGENTS

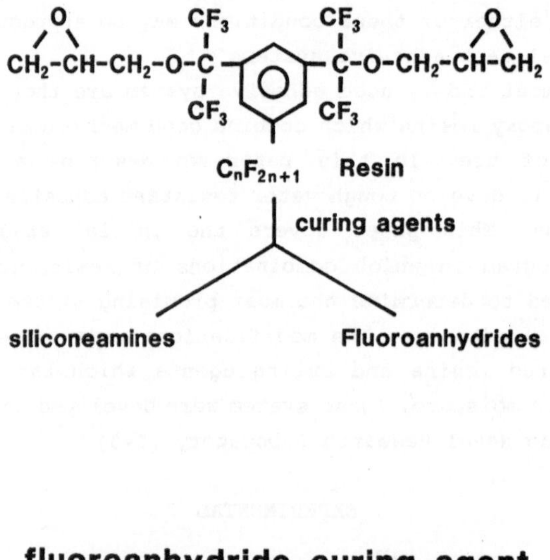

C_nF_{2n+1} Resin

curing agents

siliconeamines   Fluoroanhydrides

FA fluoroanhydride curing agent

DA dianhydride curing agent

SA silicone-amine

CAB cetyltrimethyl ammonium bromide

DMB dimethylbenzylamine

Figure 1. Fluoroepoxy resin structure and the curing agents and accelerators used

SILICONE AMINE CURING AGENTS

$$NH_2-(CH_2)_3-\underset{\underset{CH_3}{|}}{\overset{\overset{CH_3}{|}}{Si}}-\left(O-\underset{\underset{CH_3}{|}}{\overset{\overset{CH_3}{|}}{Si}}-\right)(CH_2)_3-NH_2$$

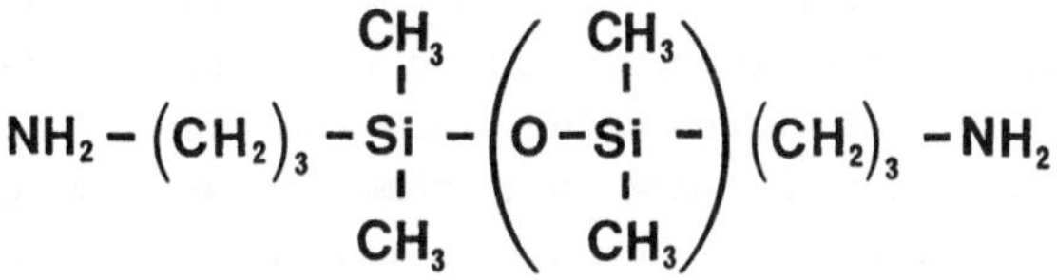

Figure 2. Silicone amine curing agents.

FLUOROANHYDRIDE CURING AGENTS

(FA)

(DA)

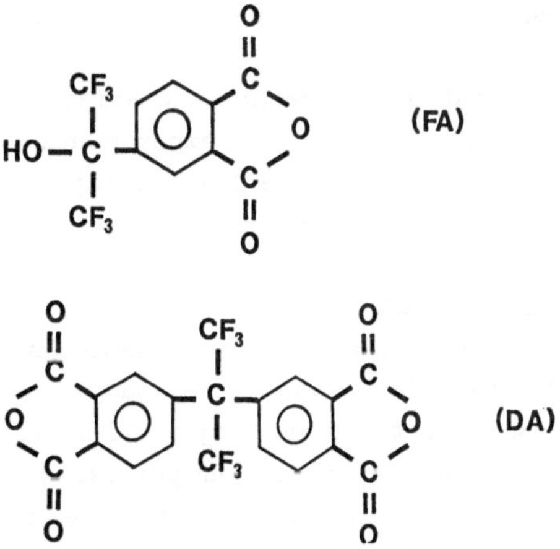

Figure 3. Fluoroanhydride curing agents.

together with catalysts to accelerate the reaction. The exact details of the mechanisms of cure of these systems has been previously reported (6). Using the various curing agents it is possible to manufacture a number of different cured resin systems and these are shown in table 1.

Water Absorption Studies

Small discs of resin (24mm diameter by 2.5 mm thick) where first died for a month using predried molecular sieve. The samples were then weighed and placed in distilled water maintained at 20C, 40C, 60C and 90C. At suitable intervals the samples were removed and after the excess water was removed were weighed on a analytical balance and returned to the bath. For each resin system and immersion temperature a number of samples were used.

Fracture and Mechanical Properties Studies

In these tests sheets of resin were cast approximately 6mm thick and machined into suitable test pieces. Room temperature fracture properties were determined using compact tension specimens (7). A sharp natural crack was introduced into these specimens before test. The specimens were tested in an Instron Tensile Testing machine at a displacement rate of 1 mm per minute. The fracture energy values were derived from the critical stress intensity factor using a Poisson's ratio of 0.35.

Values of flexural modulus, strength and strain at failure were obtained from flexural bending experiments conducted on rectangular bars according to ASTM D790-71 (8) using a crosshead displacement rate of 1 mm per minute.

Dynamic Mechanical Analysis

Samples were tested using a Rheometric Dynamic Analyzer 7700 mk2 in a torsion rectangular mode. The sample size used was approximately 6 mm thick by 50 mm long by 10 mm wide and testing was carried out at a rate of one radian per second. A sinusoidal deformation was applied to one end of the sample and the resultant torque measured at the other end. These two signals are analyzed to give G' the elastic shear modulus, G" the loss modulus and tan delta the loss angle. Measurements of these

Formulation designation	Formulation	Cure conditions hr/°C
A	$C_0/1SA$	16/20 3/110
B	$C_6/1SA$	16/20 3/110
C	$C_6/7.5SA$	16/20 3/110
D	$C_8/1SA$	16/20 3/110
E	$C_8/FA/CAB$	16/120
F	$C_8/FA/DMB$	2/75 1/100 5/150
G	$C_8/FA/DA/DMB$	2/150 2/180

Table 1. Fluoroepoxy resin/curing agent formulations and cure conditions.

values were taken automatically at a temperature resolution of
about 5C from -100C to 150C.
The glass transition temperature were determined from the maximum
in the tan delta curve. Samples of resin were also tested using
this technique after immersion in water for eight weeks at 40C.
These tests were used to determine the depression of the glass
transition temperature by the presence of water.

RESULTS AND DISCUSSION

Water Absorption Studies

Figures 4 to 7 show the moisture uptake as a function of square
root time normalised for specimen thickness for a variety of
resins systems immersed in water at 20C, 40C, 60C and 90C.
As illlustrated by the room temperature immersion trials shown
in figure 4 many of the formulations show deviations from Fickian
behaviour (9) and this is particular true for the C0/ISA and
C6/7.5SA systems. This figure demonstrates how changes in both
the resin and curing agent have a major effect upon the moisture
absorption characteristics. Generally the resins systems cured
with anhydrides have much higher water absorption values than
those cured with silicone amines. Even changes in the
accelerators used have a major effect upon moisture level. If
the C8/FA/CAB system is compared to the C8/FA/DMB the total water
absorption changes by about 50%. It is likely that the reduced
hydrophobicity exhibited by the fluoroanhydride systems could be
associated with the hydroxyl groups on the anhydride FA remaining
accessible to water molecules after crosslinking. In the silicone
amine cured systems two major points are shown. Firstly if the
fluorine content is increased then so is the hydrophobic
characteristics of the cured resin system. This is shown in
figure 4 where the three resins systems C0/1SA, C6/1SA and C8/1SA
show a direct correlation between an increase in fluorine level
and water absorbed. The second point is that increasing the size
of the silicone amine molecule produces a cured product having
reduced hydrophobicity.
In figure 5 the moisture uptake curves are shown for immersion
of samples at 40C. These curves show similar trends to those at
20C except for the C6/7.5SA resin system. It is likely that this

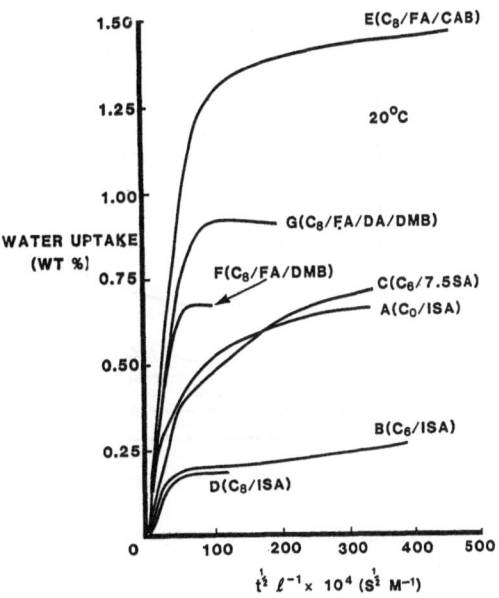

Figure 4. Water uptake of fluoroepoxy resins at 20C.

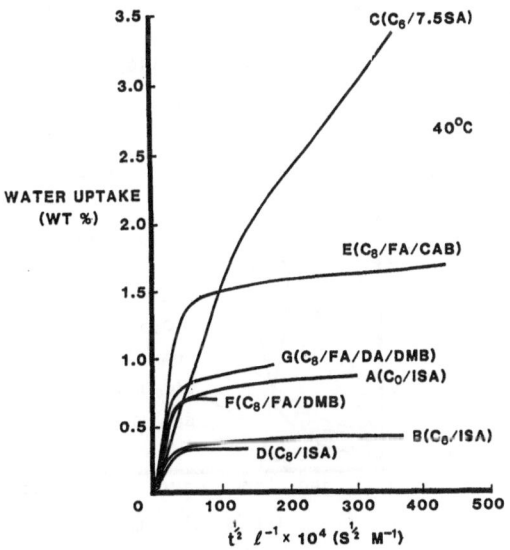

Figure 5. Water uptake of fluoroepoxy resins at 40C.

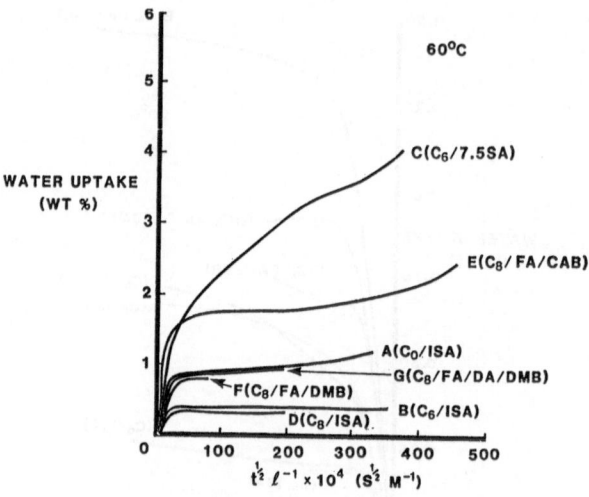

Figure 6. Water uptake of fluoroepoxy resins at 60C.

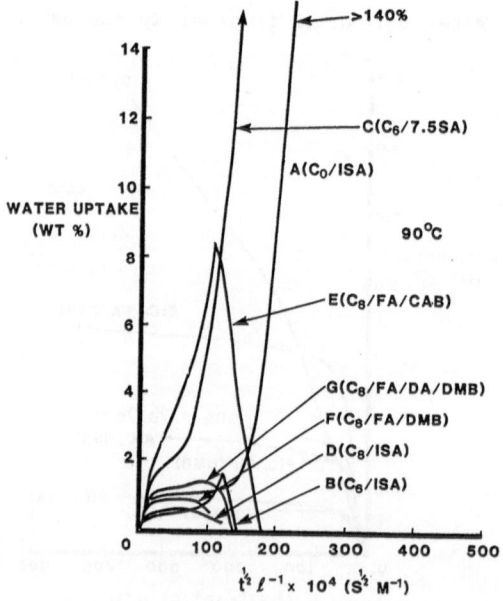

Figure 7. Water uptake of fluoroepoxy resins at 90C.

205

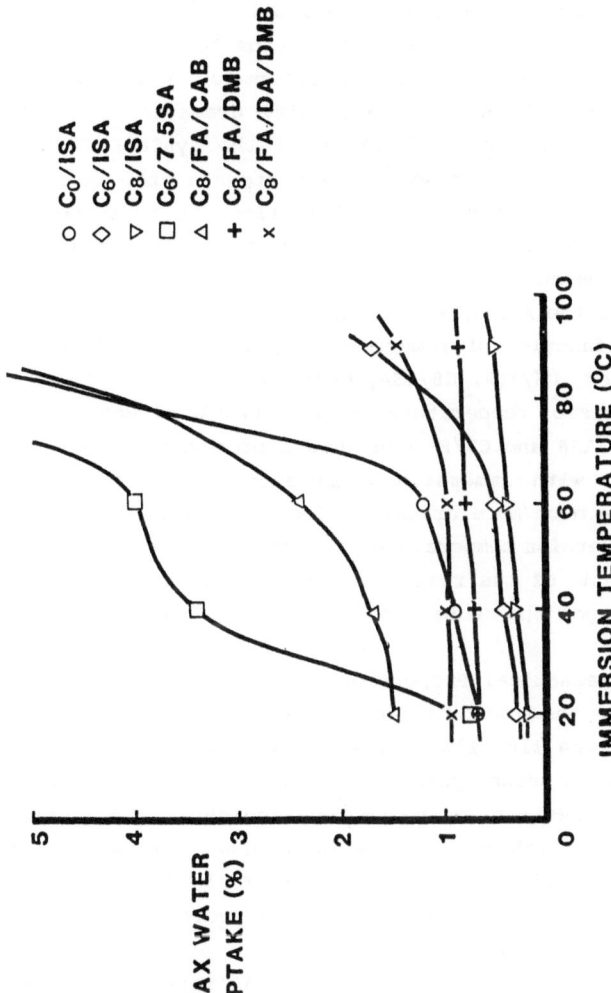

Figure 8. Maximum water uptake at various temperatures.

system has a low glass transition temperature and the immersion was at a temperature higher than this. Such a situation would allow greater amounts of water to be absorbed into the material. Figure 6 shows water absorptions curves for the resins immersed in water at 60C. These show similar trends to those immersed at 40C. In figure 7 the results are shown for a temperature of immersion of 90C. As can be seen quite dramatic increases in moisture uptake are seen for three of the systems C8/FA/CAB, C0/1SA and C6/7.5SA. In some of the systems a drop in moisture level is shown after a period of immersion. This may be due to disruption and leaching of the resins during the high immersion temperature immersion.

Figure 8 shows the maximum water uptake level for various resin systems as a function of immersion temperature. In four of the systems studied, C6/1SA, C8/1SA, C8/FA/DMB and C8/FA/DA/DMB the effect of immersion temperature has very little effect up to 60C. Two resins C0/1SA and C8/FA/CAB show a progressive increase in water content with immersion temperatures. The resin system C6/7.5SA has a much more dramatic rise in moisture content with increasing immersion temperature. Clearly the excessive moisture absorbed by some of the resins at the higher temperature has a major effect upon the mechanical integrity of the resins.

Fracture and Mechanical Properties
Table 2 shows the room temperature mechanical property data obtained from the five fluoroepoxy formulation studied. If the C6/1SA and C8/1SA resin systems are compared it can be seen that an increase in resin fluorine content indicates a reduction in modulus together with an increase in toughness in terms of both fracture toughness and fracture energy. Table 1 indicates that the type of curing agent employed has a profound influence on the mechanical behaviour of the cured product, with the anhydride-cured systems exhibiting substantially higher moduli than their equivalent silicone amine cured counterparts. These trends can clearly be attributed to major differences in structure between the two types of systems. The silicone amine curing agents would clearly impart molecular flexibility to the crosslinked network and thus influence mechanical properties in

Formulation	K_{I_c} MNm$^{-\frac{1}{2}}$	G_{I_c} Jm^{-2}	E GPa	σ_f MPa	e_f (%)	T_g (°C)
C$_6$/1SA	0.488	120.8	1.74	42.79	2.6	61
C$_8$/1SA	0.548	160.0	1.66	41.72	2.7	60
C$_8$/FA/CAB	0.426	71.1	2.24	69.07	3.6	142
C$_8$/FA/DMB	0.455	76.3	2.38	40.72	1.7	116
C$_8$/FA/DA/DMB	0.495	96.2	2.23	26.90	1.23	112

Table 2. Mechanical property values for five fluoroepoxy resin formulations

Formulation	Water Uptake	Estimate	Actual
C8/1SA	0.32%	6.4C	3.5C
C6/1SA	0.30%	6.0C	2.0C
C8/FA/CAB	1.28%	25.6C	3.5C
C8/FA/DA/DMB	0.81%	16.2C	6.5C
C8/FA/DMB	0.71%	14.2C	6.0C

Table 3. Depression of glass transition temperature due to absorbed water.

the manner described. This enhanced flexibility could also contribute to a greater toughness which could thus account for the marginally increased value of fracture toughness experienced by the silicone-amine cured systems. Values of the fracture energy derived from the fracture toughness and modulus values shows a greater apparent difference between the two cure regimes. As noted in the moisture studies section changing the accelerator used has a major effect upon the resultant properties of the resin systems. If the two systems C8/FA/CAB and C8/FA/DMB are compared there is a significant increase in the failure stress and failure strain for the DMB cured system.

Dynamic Mechanical Analysis

The dynamic mechanical responses of five of the fluoroepoxies studied are shown in figures 9 and 10. The loss angle or tan delta for the silicone amine cured systems are shown in figure 9. As can be seen there are two major transitions in these materials at about -85C and 60C. The higher temperature peak is associated with the glass transition temperature of the materials. Low temperature peaks are commonly seen in studies of amine cured non-halogenated epoxies and have been associated with crankshaft rotations of glyceryl units in the epoxy matrix (10-12). It is probable that the transitions seen in these fluoroepoxies is due to a similar mechanism.

In figure 10 the loss angle is shown for three anhydride cured materials. A similar low temperature transition occurs at about -100C as was seen with the silicone amine cured materials. Both the C8/FA/DA/DMB and C8/FA/DMB materials have a glass transition about 110C and the C8/FA/CAB has a transition at about 140C. The glass transition determined for these materials are listed in table 2. From the table it can be seen that the type of curing agent used has a major effect upon the resulting glass transition temperature. Such differences can be attributed to major differences in structure between the systems. The silicone amine curing agents would impart molecular flexibility to the crosslinked network and result in a lower glass transition temperature. This is reflected in higher fracture energy values obtained for the silicone amine cured systems.

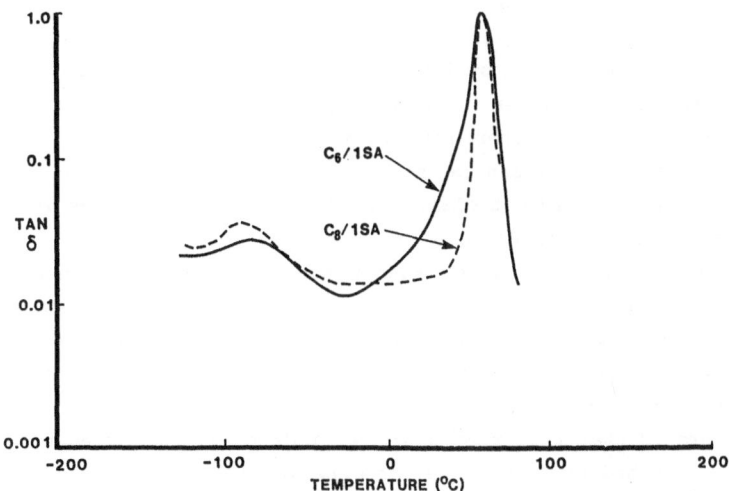

Figure 9. Dynamic Mechanical spectra for silicone amine cured
fluoroepoxy resins.

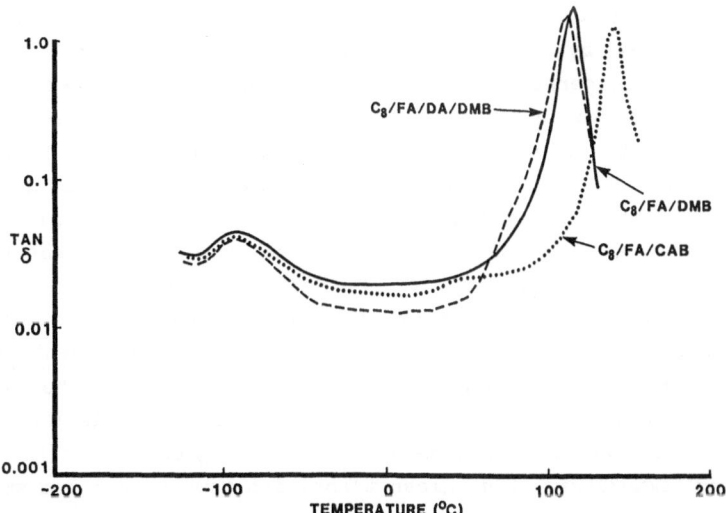

Figure 10. Dynamic Mechanical spectra for fluoroanhydride cured
fluoroepoxy resins.

The choice of accelerator used also has a major influence in the resulting glass transition temperature. In the C8/FA/DMB composition the transition is at 116C whereas in the C8/FA/CAB materials it has risen by 26C to 142C. However it should be noted that all these system were not given exactly the same cure schedule so some differences in transition temperature may be caused by this. As can be seen from tables 1 and 2 the resin system with the highest glass transition temperature was in fact cured at the lowest temperature.

A major problem with water absorption is the depression caused in the glass transition temperature. If sufficient depression is caused then a rigid structural adhesive could become quite rubberlike and therefore unable to sustain creep loads. An accepted rule of thumb for glass transition depression due to moisture is that this transition will drop by 10C for each 1% of water absorbed (13). A series of experiment were conducted where the glass transition temperature of "dry" resins was compared to those of "wet" resin systems. The wet systems were obtained by immersing suitable samples in water at 40C for eight weeks. The samples were then withdrawn and tested using dynamic mechanical spectrometry. A low temperature was chosen for immersion due to the danger of causing post curing within the resin systems. The water uptake by these samples was used to estimate a glass transition temperature depression using the factor 20C per 1% and these are shown in table 3. The actual depression measured by comparing the dry and wet dynamic mechanical response is also shown in table 3. As can be seen the real depression is significantly less in all cases than that predicted, generally the estimates are about twice the actual depressions measured. However in the case of the C8/FA/CAB material the actual depression is almost a factor of eight less that the estimate. The importance of this finding is that almost certainly the C8/FA/CAB system would have been disregarded for further work based upon its very large water uptake. In reality this material offers quite significant potential for development due to its low curing temperature, high glass transition temperature and small change in transition due to moisture. Obviously further experimental results are required before such trends can be confirmed but the results so far demonstrate the need to establish environmental changes in materials rather than estimating them.

CONCLUSIONS

Fluoroepoxy resins have been developed which when cured with silicone amines or fluoroanhydrides show substantial hydrophobicity. As the resin fluorine content is increased so is the hydrophobicity. The silicone amine cured systems have generally greater hydrophobicity than the anhydride cured materials. The glass transition temperature of the silicone amine systems is significantly lower than the anhydride cured materials. The choice of cure accelerator has a major effect upon mechanical properties and the moisture resistant of these resin systems. Some of the mechanical properties obtained from these materials are low in comparison with other non- halogenated epoxy resins.

The actual depression in the glass transition temperature due to the absorption of moisture has been shown to be significantly less than that which would normally be predicted.

REFERENCES

1   J. R. Griffith and J. E. Quick, Adv Chem Ser, 1970, **92**, 8.

2   J. R. Griffith, A. G. Sands and J. Cowling, Adv Chem Ser, 1971, **99**, 471.

3   J. R. Griffith, J. G. O'Rear and S. A. Reines, CHEMTECH, 1972, 311.

4   D. L. Hunston, J. R. Griffith and R. C. Bowers, Ind Eng Chem Prod Res Dev, 1978, **17**, 10.

5   J. R. Griffith, CHEMTECH, 1982, 290.

6   S. J. Shaw, D. A. Tod and J. R. Griffith, In Adhesives, Sealants and Coatings for Space and Harsh Environments, ed. Lieng-Huang Lee, Plenum Publishing Corporation, 1988, 45.

7.   J. F. Knott, Fundamentals of Fracture mechanics, Butterworths, London, 1973.

8.   ASTM Standard D790-71, 1973.

9.   H. Fujita, Adv Polym Sci, 1961, **3**, 1.

10. F. R. Dammont and T. K. Kwei, <u>J Polym Sci</u>, 1967, 5, 761.
11. O. Delatycki, J. C. Shaw, and J. G. Williams, <u>J Polym Sci</u>, 1969, 7, 753.
12. J. G. Williams, 1979, <u>J Appl Polym Sci</u>, 23, 3433.
13. W. W. Wright, <u>Composites</u>, 1981, 12, 201.

Copyright (C) Controller HMSO, London, 1991.

13

PROMOTING THE EXPLOITATION OF ADHESIVES IN INDUSTRY

JONATHAN WILLIAMS
Centre for Exploitation of Science and Technology
(CEST)
Enterprise House
Manchester Science Park
Manchester M15 4EN

INTRODUCTION

It is widely recognised that adhesive bonding and sealing
should play a major role in future manufacturing and
construction. This is because bonding can offer
significant potential benefits:

- Performance and sometimes cost advantages;
- Simpler design configurations and assembly
  processes;
- Possible use of new materials.

A recent market survey [1] has predicted a 10% growth in
UK consumption of high performance adhesives, with higher
growth of certain adhesive types (eg water based) and in
certain industrial sectors (eg building and
construction). However, this growth is largely restricted
to established applications in a limited number of
companies. There remains a large part of industry which
neglects the potential advantages that adhesives and
sealants might offer.

Industry as a whole is reluctant to incorporate this
technology for two very good reasons:

- Lack of confidence in long-term bond performance
  attaches significant risk to the wholesale
  introduction of adhesive materials;

- Their piecemeal introduction is often not
  cost-effective.

For example, in the field of structural silicon glazing for buildings, consulting engineers Ove Arup "... remain unassured as to the long term durability of the bond between glass and sealant." [2]

Soon after its formation CEST identified adhesive technology as an area of under-exploited potential. We began working on this field in October 1988, with the objective of examining the issues involved, and taking appropriate actions to promote exploitation.

APPROACH

The project has been divided into three phases:

- Initial assessment of the problem;
- Definition of obstacles to exploitation and ways to overcome them;
- Initiation of necessary actions.

An important feature of this work has been the high level of interaction with other interested parties: companies, research organisations, Government departments and trade bodies. This provided an objective understanding, and assured compatibility with other initiatives in this field.

Initial Assessment
The range of possible applications of adhesives and sealants is limitless. It was necessary, therefore, to restrict our attention to those industrial sectors where a significant and widespread impact could be achieved. Four sectors were selected:

- Automotive;
- Heavy engineering;
- Aerospace;
- Building and construction.

These embraced companies which had several decades of experience with adhesives, and also those with no experience; they also covered a wide range of innovation perspectives and fabrication methods.

It became clear very quickly that under-exploitation was due to a variety of causes, some technological and others organisational. Therefore, an effective response would need to involve a range of organisation types.

Obstacles to Exploitation
Two working groups were created to identify the obstacles
to exploitation and decide an appropriate set of actions.
These are illustrated in Figure 1.

The Technical Advisory Group (TAG) consisted of
representatives from adhesive user companies (from the
four selected sectors), material supply companies,
research bodies (HEI's and RA's) and Government. This
forum provided a unique interaction mechanism between
people who understood the problems of using adhesives and
sealants, and people who were knowledgeable about
potential avenues for overcoming those problems. The mix
of user companies across a range of diverse sectors
highlighted a number of generic problems; it also
indicated to what extent solutions from one sector might
be transferable to another sector.

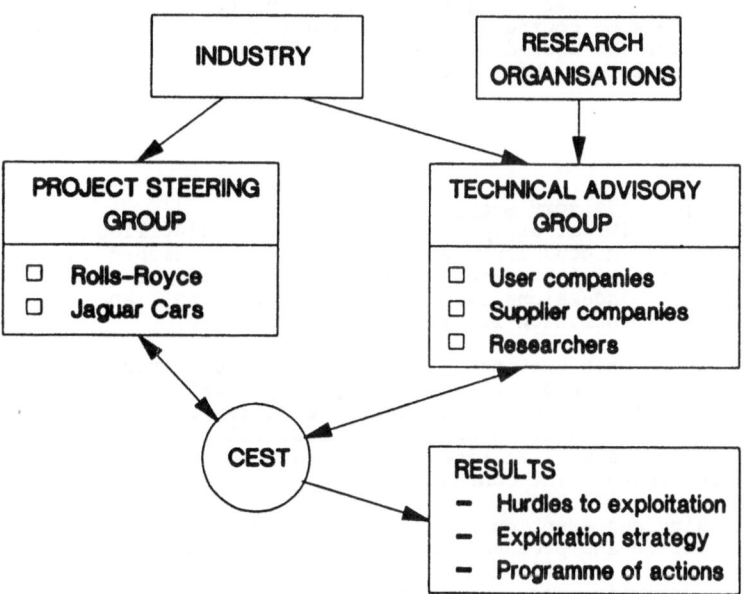

Figure 1. Working groups used to define obstacles to
exploitation and strategic response.

The Project Steering Group consisted of senior (board
level) representatives from two major manufacturing
companies, Rolls-Royce and Jaguar Cars. This peer group

was responsible for assessing the project's progress and
contributing to the strategic perspective.

In addition, four case studies were performed to examine
real or proposed applications of adhesive bonding. One
study was performed for each selected industry sector, as
follows:

-   Redesign of a windshield assembly using adhesive
    bonding, on a military aircraft;
-   Strengthening of a car suspension member using
    weld-bonding;
-   Assembly of switch-gear housings using adhesive
    bonding in place of spot welding;
-   Status of structural silicon glazing in the UK
    construction industry.

Actions

Having defined the obstacles to exploitation and a
strategy for overcoming them, it was necessary to
initiate some course of action. Apart from the fact that
they would be user-driven, we had no preconceived ideas
about what form those actions might take.

RESULTS

Technical Advisory Group

Several critical exploitation issues were distilled out
of the TAG discussions; agreement was also reached on the
types of activity needed to tackle those issues. These
detailed discussions were reported by CEST [3], and
summary information is presented in Figure 2.

It can be seen that the 'design model', consisting of
various aids to design work, plays a pivotal role in the
exploitation process. Such design aids may include
adhesive databanks to assist material selection and joint
analysis, as well as recommended practices for
establishing joint configurations.

Reaching agreement on standard data formats (both data
generation and presentation) and good design practices
necessarily constitutes a process of standardisation.
This activity must involve representatives of all
interested parties, and must be coordinated alongside
design model tasks.

Training is also essential, both to expand the skill base
and also to raise awareness of potential bonding
advantages. Currently, the majority of practicing design
engineers have an insignificant knowledge of using
adhesive materials.

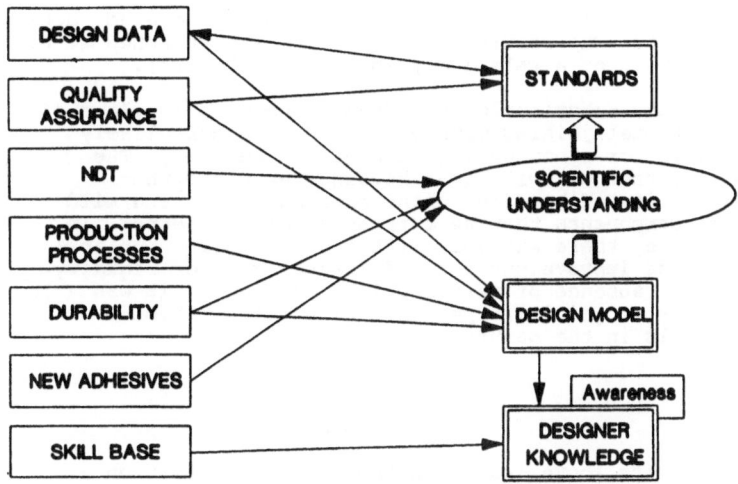

Figure 2. Exploitation issues and the necessary activity
areas.

A key result has been the establishing of basic research
as a vital element in the full exploitation process.
Several of the exploitation issues arise from inadequate
scientific knowledge of adhesion mechanisms. For example,
the ability to improve and to predict the durability of a
bonded joint requires a better understanding of bond
degradation mechanisms.

Case Studies

The case studies revealed very different attitudes to
innovation, with regard to adhesives and sealants, within
the four different industrial sectors. The principal
controlling factor, not surprisingly, was seen to be the
level of experience (or lack of it) which existed in
comparable applications.

The study of the redesign of an aircraft windshield
assembly showed how cost saving is dependent upon whether
the whole assembly is designed with adhesive bonding in
mind. In this case, bonding allowed three plastic

injection moulded components to replace ten brazed, metal
components. The resultant cost saving was reckoned to be
94% (albeit of a relatively insignificant part).

The direct substitution of adhesive bonding in place of a
conventional joining method is perhaps a more likely
scenario for a newcomer to adhesive technology. The
manufacturer of switchgear housings examined the
feasibility of bonding, using precisely the same mild
steel components that he currently welds together. In
this case, there was very little cost impact, but an
immediate improvement in performance (greater rigidity,
sealing, absense of weld pocking). Moreover, it was a low
risk way of gaining experience with using adhesive
materials in the manufacturing environment.

ACTIONS

The results above indicated the need for action on a
variety of fronts. A range of research and development
projects are required, going right back to basic research
on adhesion fundamentals. Likewise, existing and emergent
technologies must be transferred into industrial
practice, through applied development, standardisation,
training and dissemination.

It is also vital to ensure that resultant actions are
consistent with programmes and initiatives which are
already in place.

For these reasons a degree of coordination is essential,
both to achieve concurrent advances on technological
fronts, and to facilitate compatibility with third
parties. This coordination must be seen to have
technological and sectoral independence.

Centre for Adhesive Technology
A Centre for Adhesive Technology (CAT) is being created,
to stimulate and manage the necessary activities, and to
provide the level of coordination required. It will
become operational on 1st May 1990. The various functions
of the CAT are described in Figure 3.

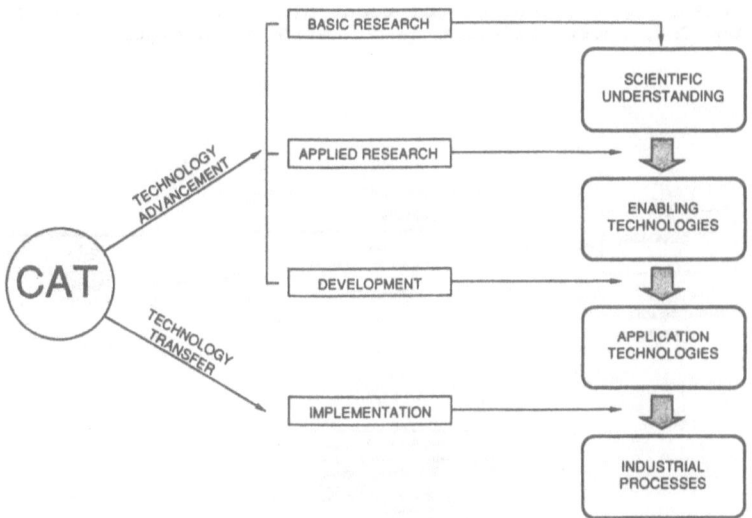

Figure 3. Principal types of activity for the CAT.

The technology transfer activities will have the most immediate impact on industrial exploitation of adhesive technology. Initial work is being planned on various design aids, including development of adhesive databank facilities to match the needs for data at the various stages of design.

The technology advancement activities will also play a vital role in creating the applicable technologies for exploitation. The CAT has a long term role, in which improved scientific understanding is translated into industrial advantage.

To be effective the CAT must retain its independence, to be sensitive to a broad range of industrial needs, and to respond with unbiased technological expediency. By locating the CAT on the site of The Welding Institute, however, advantage can be taken of the obvious synergy between these two organisations. The CAT will also display an international perspective, in respect of the user companies and centres of technical excellence with which it liaises.

Considerable attention has been paid to the management of the CAT, whose structure is illustrated in Figure 4.

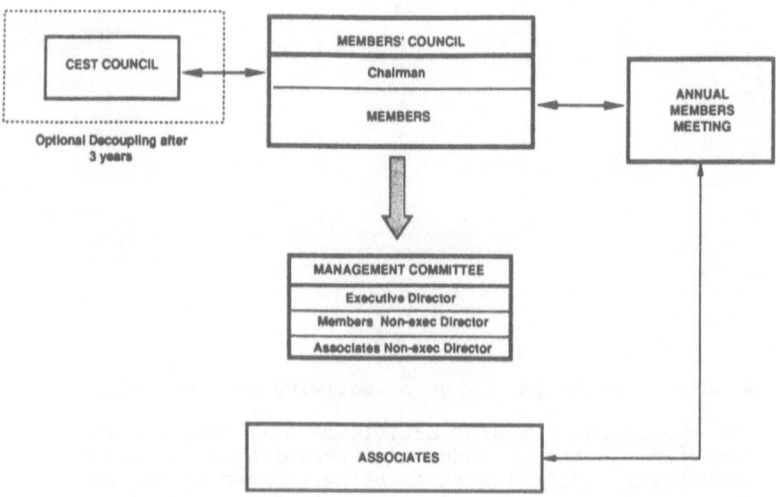

Figure 4. Management structure for the CAT.

The Members' Council is the interesting element of this structure, since it is the means of imbueing the necessary flexibility into the CAT. This Council consists of representatives from the dozen or so Member companies, who provide financial and strategic inputs to the CAT. In this way, the CAT will retain its relevance to industrial need and also develop in the best direction to respond to those needs.

The executive director has just been named as Alec Beevers, well-known for his expertise in many industrial bonding applications. He and the non-executive directors will react to the Members' Council and implement its directives. A larger group of Associate companies will receive information on CAT activities, and provide a feedback through the annual members' meeting. After the CAT is formed, CEST's only formal link will be through the Chairman, who is appointed by CEST for the first three years.

CONCLUSIONS

This project has confirmed the belief that adhesive bonding has a signifiacnt and far-reaching role to play in a wide range of industrial sectors, yet adhesive technology remains an under-exploited opportunity.

CEST has formulated and undertaken an extensive process of establishing the reasons and cures for this situation. This process has revolved around a technical advisory group (TAG) and a peer-level steering group. The constructive results from this approach have served to indicate the value of such groups. In particular, the TAG provided a unique forum for communication between a divers range of relevant organisations.

This consultative process has been reinforced by a number of case studies. This has highlighted some key advantages resulting from the introduction of adhesive technology, whilst also emphasising the need to minimise risk associated with such innovation.

Several key exploitation issues have been identified, with a strategy for tackling them. This involves coordinated action across several interdependent fronts: standardisation, research and development, design model activities and training. The need for unbiased coordination and technical independance required that a new Centre for Adhesive Technology (CAT) be formed.

The CAT is a novel type of organisation, whose objective is to promote the ability of companies to reap the opportunities offered by adhesives and sealants. It will work alongside companies, research bodies and government agencies to meet the technological needs of industry.

The ultimate objective of this initiative is to generate an international centre of excellence for joining, located in the UK. This will allow companies to improve their performance through application of the most appropriate technology.

REFERENCES

1. CTA Ltd., The UK Market for Synthetic Adhesives and Sealants - Quantitative Summary, CEST confidential report, 1989.

2. King, P., Private communication, Over Arup and Partners, 1987.

3. Williams, J.J.E., Adhesives Project Technical Advisory Group - Final Report, CEST, 1989.

14

THE DEVELOPMENT AND TRANSFER OF ADHESIVES TECHNOLOGY AT PERA

JOHN HILL
Manager, Polymer Technology Group,
PERA, Melton Mowbray, Leics. LE13 0PB

1. ABSTRACT

PERA the Melton Mowbray based Research and Technology Centre, have displayed an ongoing commitment to increasing the exploitation of adhesives bonding within the UK and European industry.

Through their close links with industry and a series of surveys amongst current and potential end users, PERA have identified the three main problems in encouraging industry to exploit fully the benefits of this joining medium

Companies currently lack the confidence to exploit adhesives successfully due to :-

DESIGN	- the absence of accessible, user friendly design theory, procedures and software.
MANUFACTURING	- the absence of reliable, quality validated automation technology to handle high volume products.
RELIABILITY	- the absence of any form of on-line, real time NDT for joints.

PERA's impressive programme of technology development and research recently completed, is to be consolidated with four new projects and a comprehensive Europe-wide dissemination programme.

PERA's most recently completed projects focus their 10 years of background research in the areas of :-

DESIGN	- ADENG. The development of design procedures, software and databases for general engineering applications.
MANUFACTURING	- ADPROC. The development of an automation application technology for bonded assemblies, including the

on-line validation of adherends and
beam form.

The new projects will complete the development and transfer of this
essential technology to industry.

DESIGN   - ADENG II. The development of design
procedures, software and databases for
high performance, high durability
applications.

- ADCAD. The development of a dedicated
adhesives CAD system to complement
currently available CAD hardware and
software.

MANUFACTURE   - ADPROC II. The completion of the fully
automated/validated bonding concept
with on-line surface preparation, the
quantification and identification of
adherend contamination, and the closed
looping of bead form production

RELIABILITY   - BoNDTest. The development of on-line
NDT equipment able to respond in real
time within production environments.

By the mid 1990's PERA's current initiatives will have provided
European industry with the tools to exploit adhesives bonding
successfully on a scale likely to generate a major competitive edge
for the community as a whole.

Further information on the completed projects shows the potential
advantages of automatic control of adhesive processing allied to
automatic non destructive inspection with particular reference to
improvement and consistency in quality of production. Other
potential advantages, such as consistent and predictable output
levels, reduction in direct labour content, and lowering of
exposure of personnel to health and safety hazards, are also
readily apparent.

With respect to the design aspects of PERA's work creating adhesive
joints for engineering applications is no longer an art form
calling for a trial and error approach with necessarily high
margins of error.

The information available, and the way in which it can be
presented, can now give the designer the knowledge that he has the
facilities to devise a joint based on readily available
scientifically substantiated data, and from this knowledge, the
confidence to design with a high level of precision, joints to meet
the exacting requirements of current industrial and commercial
demands.

2. COMPLETED WORK ON DESIGN ASPECTS OF ADHESIVE JOINING

The potential advantages from the use of adhesives for engineering
applications have been realised for many years. These include the
aesthetic value of smooth surfaces unbroken by bolts, rivets or
spot welds, the simultaneous sealing and bonding of a joint, the
elimination of accurately machined bolt holes and the spreading of
the load over the entire joint surface, thus reducing local
stresses etc.

Unfortunately, whilst a considerable amount of good theoretical
work has been done on the stresses, etc. of bonded joints, an
accepted design procedure, suitable for use by design engineers,
was not available and the repeatability and life of joints was very
suspect.

The manufacturers of adhesives were willing to supply data on their
products, but did not know exactly what was required by the
designer.

The objectives of the ADENG project were to overcome these
difficulties by developing a design procedure, determining which
adhesive properties were required by the designer and suitable test
procedures and also the most suitable surface preparations to
ensure repeatable and long joint life.

As a result of the four years work on the project, the designer now
has design software with a procedure manual, thus enabling him to
select the correct joint type and size. He also has data bases
which enable him to select and source the correct adhesive and the
best surface preparation for optimum joint life and reliability.

2.1 Overall Objective

The aim of the project was to seek research and performance data
and develop the basic technology to enable companies in the
engineering sectors of European industry to take advantage of the
unique benefits of adhesive bonding technology.

Comprehensive guidelines were to be developed and evaluated, which
would enable engineers to design, manufacture and control the
quality of adhesive bonded assemblies.

2.2 Overall Conclusions

It was originally envisaged that it would be possible to develop a
design procedure for adhesively bonded joints which could be
described as 'rule of thumb'. The ADENG project has, however,
resulted in a detailed procedure which uses computer software to
obtain accurate design information for adhesively bonded joints.

Whilst the strength of bonded joints is of paramount importance,
they must also be of repeatable quality and have a satisfactory

life in the ambient environmental conditions. Data bases have been developed which enable the correct adhesive type to be selected and sourced. Further data bases list appropriate European Design Standards and the surface preparation to be chosen to meet the quality requirements.

2.3 Joint Configuration Study

The objective of this task was to define, in depth, the types of joint required by industry and the modes of applied load, whilst assessing their relative importance.

The study resulted in the selection of ten different configurations of joint, each of which could be loaded in a number of ways, such as tension, compression, bending, shear or torsion.

This resulted in a total of 28 configurations which were then placed in a general order of importance. From this it was determined that the most important joint types were the single lap joint, the double lap joint and the cylindrical joint.

2.4 Design Study Elastic and Plastic Modes

The aim of this task was to develop methods for the design of joints in which the components of the joint are loaded both within and beyond their elastic limits. Initially the work on the elastic and the plastic modes was carried out simultaneously, but as two separate tasks. Eventually, however, the two modes were considered as one study.

Technology reviews and research into techniques for joint performance predication led to the development of equations and preliminary elastic design software. This was initially for lap joints, but later for most of the joints selected. After practical testing of joint configurations, a design procedure was outlined, and initial software developed, for both the elastic and plastic modes. Finally, after more detailed consideration of the yield characteristics of the adhesive, the methodology was refined and software produced with an accompanying design manual.

As a result of this work, engineering designers will now be able to incorporate adhesively bonded joints into their structures, confident that their calculations have been based on a proven methodology.

2.5 Adhesive Selection Programme

The objective of this task was to establish a data base of engineering adhesives using microcomputer software, enabling requirements to be matched with available adhesives and so aid adhesive selection.

When selecting an adhesive for a specific application it is necessary to consider the service requirements such as operating temperature etc., health and safety, compatibility of the curing regime with other processes, availability, cost, surface preparation and conformation to standards.

A data base has been constructed which makes use of three computer programmes which contain information on standards, manufactures, surface pretreatment and adhesive properties. It is menu driven and easy to use with the results available from either the screen or printer. This enables the designer to readily select the most suitable adhesive for a specific application whilst taking into consideration all the relevant constraints.

2.6 Computer Aided Design

It was considered at the outset of this project that a logical step would be to incorporate into a CAD program, the design procedure developed during the ADENG project. The objective of this task was to investigate ways of achieving this. However, after some work and discussion between the three partners, it was considered that the time allowed for this task would prove to be inadequate for the amount of work necessary. It was therefore decided that a report should be written which would explain what had to be incorporated into a CAD programme and any technical difficulties, and propose a plan of work which could be carried out in the future to enable the original objective to be achieved. Any unused time on this task was reallocated to the design study and adhesive performance tasks.

The results of this task have shown that although there are likely to be some difficulties encountered with respect to compatibility of software, further work outside the project could lead to the incorporation of the results of the ADENG project into a CAD system.

2.7 Adhesive Performance

The objective of this task was to produce detailed data on the behaviour, as engineering materials, of up to ten adhesives and to investigate the effects of production variables on joint performance. It was also necessary to develop tests which would yield suitable data for design purposes.

Ten adhesives from a range of types were selected for testing. It was accepted that the simple lap shear test results usually supplied by manufacturers of adhesives were not the properties required by the designer. After extensive trials the thick adherend shear test was selected as being the best test for the provision of the required shear modulus and yield properties.

The selected adhesives were then tested in this way to obtain their mechanical properties. Tensile testing was also carried out upon bulk specimens of the adhesive and good correlation was achieved between these and the thick adherend shear tests.

The effect of water absorption upon the glass transition temperature was determined using a torsion pendulum technique and also a series of load cycling trials were made to attempt to clarify the difficulty of determining the yield or 'limit of reversibility' stress for an adhesive.

This work has enabled a test procedure to be developed which enables the required mechanical properties of an adhesive to be determined with confidence in their reliability and repeatability.

Further work has also generated a large amount of information on the mechanical properties of a number of adhesives covering a range of chemical types.

2.8 Surface Pretreatment Processes

The objectives of this task were to select a range of materials and potential surface treatments for evaluation, to develop experimental techniques, such as accelerated ageing, and compare them with natural service environment conditions. Finally, after extensive trials, specific recommendations were to be made for pretreatment methods and processes on engineering materials.

Primary and secondary lists were developed of eight metals and six polymers with up to four treatments identified for each material. Condensing humidity conditions on small lap joints were found to give reliable results with accelerated ageing tests, and from this work, durability test procedures were developed using perforated lap shear joints.

Extended environmental tests and durability trials were carried out using roof top sites, test panels suspended below road vehicles and a polymer/metal assembly subjected to natural weathering with low frequency load cycling.

Durability and reliability of bonded joints have always been subjects of concern to potential users of adhesives in engineering fields and this work does much to remove the problem by giving definite recommendations for surface treatments after proving their efficiency by both laboratory and service trials.

2.9 Adhesive Application Techniques

Adhesives can have rheologies ranging from free flowing liquids to thick pastes and the objectives of this task were to investigate and review equipment and methods which enable their controlled application to engineering components. Surveys were carried out, and reports prepared, upon the types of equipment available for both the dispensing and application of adhesives of all types. Typical applications were also given for several different types of joint where such equipment as robots and vision systems were employed.

One of the main advantages of adhesives in engineering is the
possibility of semi or total automation of the process, with the
obvious advantages of speed and process control and this work
enables the design engineer to appreciate the current state of the
art in this field.

2.10 Component Assembly Methods

The objective of this task was to investigate methods and equipment
that might ease the assembly and fixturing of components to be
adhesively bonded.

A study was made of the requirements of bonded joints as compared
to the standard assembly of components with regard to both plane
and circular joints. The possibility of post assembly adhesive
application was also the subject of an investigation.

As a result of this work and extensive searches of published
literature it was concluded that whilst some information was
available upon the assembly of specific items, it was not possible
to generalise upon the overall subject of assembly.

It was considered necessary for the assembly method to be covered
in detail during the design of every product in the same way that
assembly of non-adhesive joints is normally considered.

2.11 Quality Control of Joints

The quality control of joints can be considered to consist of
detection of the presence of adhesive on components before assembly
and non-destructive testing of finished joints for integrity. The
objective of this task was to investigate, practically compare and
evaluate methods of carrying out this quality control.

A number of systems were found to be available which will detect
the presence or absence of adhesive, but many of them were somewhat
insensitive and could at best detect only the absence of large
quantities of adhesive and would overlook small discontinuities.
Vision systems on the other hand were found capable of detecting
not only the presence of the adhesive, but also whether there was
the correct amount.

Many techniques were found to be available for the non-destructive
testing of joints, including several forms of ultrasonic testing,
radiography, holography and acoustics.

Each of the processes has value and different types of defect and
joint type call for different test techniques, but the work has
produced an overview of quality control which enables the design
engineer to approach adhesive bonding with greater confidence in
the final product.

2.12 The Application and Proving of the Technology

It was the object of this task to apply, prove and refine the
technology developed during the ADENG project by designing,
manufacturing and testing bonded structures and engineering
components.

Four case studies were conducted in this task which consisted of an
intermediate shaft of the type which would be used, for instance,
to connect a pump to an electric motor, an aluminium panel with a
stiffening spine bonded across the middle, a compressor crankshaft
and a compressor crank case. A number of different adherend
materials and adhesives were used, in each case with manual methods
of assembly and adhesive application. The final components were
all subjected to mechanical testing, in some cases at different
temperatures, with a view to comparing the strains produced by
applied loads and the yield stresses of the joints with the
theoretical results calculated when using the design procedure.

The results of the tests indicated a good relationship between
practice and theory and engendered a good order of confidence in
the technology of adhesive bonding which was developed in the
project.

3. **SUMMARY OF COMPLETED WORK ON PROCESS DEVELOPMENT
ASPECTS OF ADHESIVE JOINING**

The joining of materials using adhesives has been known for
thousands of years. As a manufacturing process the use of adhesive
bonding has been limited by the relatively slow speed of
application and by uncertainty as to the efficiency of the bond.

The demand for low weight structures in the aircraft industry led
to the extensive use of adhesives under very tightly controlled
conditions, with inspection and testing at every stage. The
benefits of lightness and the ability to join dissimilar materials
without mechanical fasteners outweigh the disadvantages of relative
slowness and necessity for frequent testing, since these factors
are not a major problem in an industry where numbers produced are
not high.

In order that the benefits of adhesive joining could be generally
applied to high volume production consumer items - white goods,
automotive production - a major technological breakthrough has been
long awaited. The main hope has been that a non-destructive
inspection method would be developed which would reliably determine
the strength of an adhesive bonded joint, thereby ensuring that
such a joint could bear the load for which it was designed.

The breakthrough has not been forthcoming, and after a number of
years its arrival appears more and more unlikely. From time to
time there are reports that a specific configuration, under tightly
controlled laboratory conditions can be evaluated, but the
development of a production tool to determine bond strength
non-destructively does not appear to be close.

To have reasonable assurance of the quality of an adhesive bonded
joint in the absence of such a development, it is necessary to
control rigidly each stage of the production cycle, with
appropriate testing and inspection before proceeding to the next
stage. In order to obtain this assurance it is necessary to be
reasonably certain that :-

a) The Material to be joined is clean.

Adhesive joining involves molecular bonding of surfaces. In an
ideal case this means that each of the joined surfaces will have a
molecular bond with the layer of adhesive between them, and that
the strength of these molecular bonds is known. If, however,
foreign matter is present on one of the surfaces prior to joining,
for example oil, or an oxide layer, then the molecular bonds in
that area will be first adhesive to foreign matter and second
foreign matter to surface. The strength of such bonds is unlikely
to be known, particularly where the nature and quantity of the
foreign matter is unknown.

b) The correct amount of Adhesive has been applied to the correct
    position.

The basis of joint design is that a known area of known materials
will be bonded by a known thickness of correctly cured adhesive.
In order to develop the desired properties in the joint it is
important that the adhesive layer should be of the correct
thickness. The volume applied should therefore be sufficient to
produce a layer of the specified area and thickness, and should be
in the correct position to achieve this.

c) The presence is known of any Physical Defect or Condition which
    could Affect the Properties of the joint.

Such defects as voids, unbonded areas, included foreign matter, if
present in a joint, if they were sufficiently large, or at a
particularly high stressed position, could materially effect the
properties of a joint, to the extent that they could render it
unacceptable for service conditions.

The presence and position of these defects must be able to be
readily ascertained.

d) Adhesive has been Applied in the Correct Condition and Cured in
    Accordance with Manufacturers'/Joint Designers Recommendations.

This aspect of adhesive bonding can only be assured by rigid
process control, covering storage, handling, and curing conditions.
Without this assurance it is possible that the final properties of
the adhesive layer may not be these upon which the joint design was
based.

As no non-destructive test has been developed which will reliably

determine the properties of an adhesive bonded joint it is apparent
that the points noted above had to be approached separately. It
was felt that if factors a, b, c could be controlled, the condition
of the adhesive layer could be assured by rigid process control,
involving identification and comprehensive records covering
receipt, storage, handling and control of curing conditions, there
would then be a realistic possibility that adhesive bonding could
be used to achieve consistent quality joints in high volume
automated production.

PERA Project

A development project led by PERA involving a consortium of
industrial sponsors, with further financial support from the
Department of Trade and Industry has investigated the factors
involved in the production of adhesive bonded joints and the
possibilities of automating production in order to assess the
potential for an automated cell in a production environment to
produce bonded joints of consistent quality.

The Investigation

Certain features were identified as being necessary to produce
consistent joints by automated processing. These were :-

a) Real time automatic monitoring of the freedom from
contamination of the areas to be bonded.

b) Automatic placing of the correct quantity of adhesive.

c) Real time confirmation that the adhesive had been placed
correctly and in the correct quantity.

d) Automatic assembly of the parts to be joined.

e) Automatic non destructive inspection in real time of the
completed joint to detect any physical defect which might
materially affect the joint's performance.

The investigation of each of these factors followed a similar
pattern - search of published literature, contact with possible
suppliers to identify any commercial available equipment, a
theoretical evaluation of any methods identified, and a practical
evaluation of the methods which were considered most likely to meet
the requirements. Where no suitable methods or equipment was
found, development was undertaken in an attempt to produce one. An
outline is given of the investigation and findings with regard to
the following subject :

1. Surface Contamination
2. Adhesive Application/Assembly of Parts
3. Confirmation of Application
4. Non-destructive Inspect of Bonded Joints.

Detection of Surface Contamination

A literature search revealed a number of methods for detecting the presence of surface contamination.

These varied from the "breaking" of rinse water film on chemically cleaned surfaces, to a number of sophisticated procedures, involving spectroscopy and x-ray emission. The use of ellipsometry is also recorded, but this requires a knowledge of the optical properties of the layer of contaminant. The following three commercially available equipments for the detection of surface contamination were identified.

1. The Fokker Contamination Tester

The presence of a layer of contaminant on the surface of a material will increase the energy required to cause electrons to leave that surface. The electron emission energy or surface potential difference of an electrically conducting material can be measured by Kelvin's dynamic condenser method.

A gold plated reference electrode, connected by a resistor to the material to be surveyed, is placed close to the surface. The Electrical field is then created in the air gap between the reference plate and the surface. When the reference plate is moved relative to the surface a charge is displaced through the resistor giving an indication of the surface potential difference. This principle is used in the Fokker Contamination Tester.

The Photo Acoustic Technology Surface Contamination Detector

When a surface is irradiated by ultra violet light, electrons are caused to be emitted from it. This is known as photoelectron emission, or optically stimulated electron emission, and is the principle which is employed in the PAT surface contamination detector. Each material has its own particular emission rate, and metals have a particularly high emission rate. Surfaces are subjected to a fixed level of irradiation and the electrons emitted are attracted to a biased collector which measures their presence as an electric current.

If a contaminant is present on the irradiated surface, the electrons emitted will be at a rate peculiar to the contaminating material and not to that of the substrate, indicating the presence of contaminant by a change in the measured current.

The Tencor Contamination Detector

this instrument operates on the principle that the presence of contamination on a surface will affect the amount of light which it reflects. The surface to be tested must be reflective and extremely smooth. If there is any surface roughness or transparency of the material the instrument will not function.

In operation a beam of laser light is projected on to the surface
and moved over it. The presence of contamination is indicated by
an increase in the amount of light which is scattered.

Assessment

The methods based on light reflection and surface potential
difference measurement had serious limitations for use in an
automated production environment. The former required a smooth
opaque surface, the latter a conductive material. Since one of the
many varied materials can be adhesively bonded, it was considered
that the method which had least potential limitations should be
practically assessed. On assessment of the PAT Surface
Contamination detector it was concluded that contaminants such as
oil, fingerprints, silicones could be readily detected on metals
and polymers, and that the equipment reacted sufficiently rapidly
to be acceptable for use in an automated survey process.

At a later stage in the investigation the PAT contamination
defector was built into an automated cell which assembled frames
and covers. After adhesive was applied to the frame a robot was
programmed to pick up a cover from a roller conveyor line. The
cover was then passed over the probe of the PAT scanner at a fixed
distance and speed, so that the whole of the surface to be
adhesively joined was scanned. A computer was programmed to give a
visual indication of level of electron emission in the scanned area
and could be set to accept or reject given emission levels, or to
alert a supervisor where a judgement of acceptability might be
required.

Automated Application of Adhesives

The main requirements in application of adhesive over an area is
that there should be relative movement of applicator and surface.
Other important requirements are that the correct amount of
adhesive must be delivered from the applicator, and in order to
form an even 'bead' the adhesive must be delivered at a rate
corresponding to the linear speed at which the applicator moves
relative to the surface. If it is necessary for the rate of
movement to be less as, for example, when negotiating a difficult
corner, then obviously adhesive will be required to be delivered
more slowly in order to maintain an even bead size.

In manual application such adjustments will be made by the skilled
operator, but in automatic application, where rapid and consistent
relative movement is necessary, a robot is employed. This can be
an articulated arm with a number of joints as required, or a
cartesian robot, X Y (Z), depending on the complexity of the
surfaces to be joined, the desired accuracy of the bead laying, and
the loads to be moved.

In order to achieve the desired conditions, in addition to being
programmable to produce the required route of adhesive application,

the robot should be able to communicate with the adhesive supply
system so that applicator speed and adhesive delivery rate can be
matched.

Virtually all adhesive used in quantity production is in liquid
form, and is usually fed to the applicator by a pump.

In order to deliver a correct amount of liquid adhesive to each
workpiece, some form of metering pump is desirable. In the case of
two part adhesives this would be a twin metering pump. The rate of
pump output would be controlled by the robot controller.

Confirmation of Bead Laying

If a robot applying adhesive to a surface follows the prescribed
motions at the required speed, while the metering pump(s) deliver
the required quantity of adhesive at the correct rate corresponding
to the speed of travel of the applicator, it is possible to assume
that the bead will be laid correctly. However, unforeseen
situations can arise to prevent the planned application such as
bubbles in the adhesive train, variations in pumping rate,
temperature changes affecting adhesive viscosity, so that a bead of
incorrect size, or even no bead at all, is laid in place. If
undetected this will ultimately lead to the production of a
substandard joint.

In manual application of adhesive, such a situation would be
observed by the operator, and where necessary suitable corrective
action would be initiated, or the affected workpiece would be
prevented from continuing through the production cycle. It is
therefore essential that a similar independent confirmation of bead
quality is available if acceptable joints are to be produced
consistently by automatic adhesive application.

Numerous devices such as ultrasonic transceivers, pneumatic
reaction cells or photo electric cells can be used close to the
orifice of the applicator to determine whether adhesive is flowing.
These devices have individual drawbacks such as size, rate of
reaction, effect of adhesive stream on them, but the major obstacle
to their use is that they do not detect abnormal flow.

Systems which can not only detect the presence of adhesive bead but
can also measure its size are grouped under the heading of vision
systems. The basic requirements of such a system are a light
source to illuminate the subject, a lens to receive the reflected
light and focus the image on to a detector, and an electronic
system to process the signals from the detector.

The main difference between available types of vision system is the
electronic processing system, although the use of laser light is
reported not only to provide information on bead thickness, but
also on adhesive bead height.

A commercially available laser beam vision system which had

originally been developed to control the deposition of weld metal
was found to have excellent resolution. However, its scanning
rate, which was perfectly adequate for speeds of weld metal
deposition, was not sufficiently high to accommodate the relatively
high rates of movement sought in automatic adhesive deposition. It
became apparent that in use such a system would not necessarily
detect defective or missing areas of adhesive bead up to 20mm long.
Since this was unacceptable and no alternative commercially
available systems were found which could cover the bead laying, the
development of a vision system which would meet the specific
requirements of automatic adhesive bead laying was put in hand.

The development was based on the use of an optical lens with a
charge coupled device (CCD) array mounted in its focal place. The
final version of the development the light which illuminated the
adhesive bead was transmitted via fibre optic bundles, and the
image was transmitted via a coherent fibre optic bundle. The main
objective in using the light transmission systems was to avoid
mounting the optical and vision processing system near to the end
of the robot arm, which would have effectively reduced the robot's
useful load carrying capacity and/or its speed of operation.

With suitable electronic processing using a single board computer,
the system was sufficiently flexible to deal with difficult
coloured adhesives on a variety of surfaces. It was found to be
possible to specify detection of particular requirements for the
adhesive bead such as minimum bead width and bead continuity. At
speeds of 400mm/sec breaks of 0.1mm in length could be detected in
a 1mm wide adhesive bead. The system could readily connected to
the robot controller so that appropriate action could be taken.

Non Destructive Inspection

Although no destructive evaluation of individual bonded joint
strength is not generally available, non destructive inspection
still has a major role to play in providing assurance that a joint
does not contain any physical features which might reduce its
overall strength. Such features would include voids in the bead,
included matter, and unbonded surfaces where adhesive and substrate
are in close contact but no molecular bond has been produced.

Various non destructive methods have been used for defect detection
with varying degrees of success, but for inspection of
automatically produced adhesively bonded structures, it is of
paramount importance that the inspection method should be capable
of operating in real time. With this in mind several methods were
investigated, including ultrasonic echo techniques, ultrasonic
resonance, acoustic emission, thermography, holographny, tapometer,
x-rays, newton radiography and vibration analysis. Of these, to
detect disbonds, voids and inclusions, it was concluded that only
ultrasonic pulse-echo, and after suitable development, vibration
analysis would be suitable. Methods such as thermography,
holography, ultrasonic resonance, acoustic emission and tapometer
are claimed to differentiate between "good and bad" bonds, but do

not necessarily indicate the position, type and size of defect.

With respect to the potential use of ultrasonic scanning techniques to inspect automatically produced adhesive bonded joints there appeared to be disadvantages in that some form of couplant is necessary between the transducer and the surface of the piece to be inspected, and that scanning of large areas could be time consuming. Attention was therefore focused on vibration analysis as having most potential to meet the requirements of automatic real time inspection.

Vibration Analysis

The method was based on the tapping technique in which the subject area is given a sharp tap, and because vibrational characteristics are affected by discontinuities the audible vibrations from a defective component give a sound which the human ear is recognisably different to that emitted by a component without defects.

Using a vibrator at a known frequency continuously striking with a known energy at various positions along one face of a bonded joint, readings of the vibrations emitted were recorded by a sensitive acoustic microphone. The signals were analysed by a spectrum analyser which gave a visual display of the vibration spectrum over the range 0-10KHz. The information was also stored in a microcomputer. Subsequent analysis showed that deliberately introduced voids, inclusions and disbonds could be detected and that with further development it should be possible to adapt the automatic tapping system to real time inspection. To establish the differences between classes of discontinuity would also require further work.

Conclusions

An automated adhesive assembly cell was constructed in which separate features pertinent to the assurance of quality of bonded joints were automated and, where applicable, included in the cell.

The relevant features, surface validation prior to bonding, accurate dispensing of adhesives, validation of adhesive bead integrity, handling of adherend before and after assembly, and non-destructive inspection of finished joints were all found to be able to be automated to a production rate compatible with the manufacture of consumer items.

Further development is necessary in certain areas, particularly non destructive inspection, but there appears to be no major technical obstacle to the construction of the manufacturing cell which will deal automatically in real time with all the factors detailed above, providing an assurance of consistency of manufacture and of sustainable quality.